Disclaimer

The publisher of this book is by no way associated with the National Institute of Standards and Technology (NIST). The NIST did not publish this book. It was published by 50 page publications under the public domain license.

50 Page Publications.

Book Title: Helium Dispersion in an Attached Single-Car Residential Garage with and Without Vehicle

Book Author: William M. Pitts; Jiann C. Yang; Marco G. Fernandez; Kuldeep R. Prasad;

Book Abstract: The dispersion and loss of helium inside a single-car residential garage attached to a single-family house was experimentally characterized by recording time-resolved helium concentrations at multiple locations in the garage and at a single location in the house during and following helium releases near the floor of the garage. Helium served as a surrogate for hydrogen for safety reasons, and helium release rates were adjusted to provide the same constant volume flow rate as that required to release 5 kg of hydrogen over a four hour period. Supporting measurements included compartment leakage, temperature, and atmospheric wind conditions. Helium was released upwards either as momentum- or buoyancydominated flows. Experiments were performed with the garage empty or with one of two conventional mid-sized automobiles parked over the release location. Six tests with the garage naturally ventilated and six tests employing forced ventilation with a fan are described. A variety of parameters were used to characterize the mixing behavior. Conclusions emphasized include: a) the role of Froude number on helium mixing behavior, b) the development of upper and lower helium concentration layers in the garage during a release, c) the measurable, but limited, effects of atmospheric wind on the results, d) the relatively efficient transfer of helium from the garage into the house during the releases, e) the ability of a vehicle to trap a high helium concentration in the engine compartment and, particularly, the undercarriage during a helium release and the relatively rapid drop in these levels to those of the surrounding garage at the end of the release, f) the relatively slow buildup of helium in the passenger compartment and trunk of a vehicle during a helium release and subsequent slow decay following cessation of the flow, g) the effectiveness of active ventilation in reducing helium concentrations in the garage to levels below those corresponding to flammable concentrations of hydrogen, and h) the trapping of helium/air mixtures corresponding to highly flammable hydrogen mixtures inside the vehicles even when active garage
ventilation was employed.

Citation: NIST TN - 1765

Keywords: attached residential garages, automobiles, buoyancy-dominated flow, forced ventilation, Froude number, helium concentration, hydrogen concentration, momentum-dominated flow, natural ventilation, wind effects

NIST Technical Note 1731

Helium Dispersion in an Attached Single-Car Residential Garage with and Without Vehicle

William M. Pitts
Jiann C. Yang
Marco G. Fernandez
Kuldeep Prasad

http://dx.doi.org/10.6028/NIST.TN.1731

National Institute of
Standards and Technology
U.S. Department of Commerce

NIST Technical Note 1731

Helium Dispersion in an Attached Single-Car Residential Garage with and Without Vehicle

William M. Pitts
Jiann C. Yang
Marco G. Fernandez
Kuldeep Prasad
Engineering Laboratory

http://dx.doi.org/10.6028/NIST.TN.1731

September 2012

U.S. Department of Commerce
Rebecca Blank, Acting Secretary

National Institute of Standards and Technology
Patrick D. Gallagher, Under Secretary of Commerce for Standards and Technology and Director

National Institute of Standards and Technology Technical Note 1731
Natl. Inst. Stand. Technol. Tech. Note 1731, 98 pages (January 2012)
http://dx.doi.org/10.6028/NIST.TN.1731
CODEN: NTNOEF

Abstract

The dispersion and loss of helium inside a single-car residential garage attached to a single-family house was experimentally characterized by recording time-resolved helium concentrations at multiple locations in the garage and at a single location in the house during and following helium releases near the floor of the garage. Helium served as a surrogate for hydrogen for safety reasons, and helium release rates were adjusted to provide the same constant volume flow rate as that required to release 5 kg of hydrogen over a four hour period. Supporting measurements included compartment leakage, temperature, and atmospheric wind conditions. Helium was released upwards either as momentum- or buoyancy-dominated flows. Experiments were performed with the garage empty or with one of two conventional mid-sized automobiles parked over the release location. Six tests with the garage naturally ventilated and six tests employing forced ventilation with a fan are described. A variety of parameters were used to characterize the mixing behavior. Conclusions emphasized include: a) the role of Froude number on helium mixing behavior, b) the development of upper and lower helium concentration layers in the garage during a release, c) the measurable, but limited, effects of atmospheric wind on the results, d) the relatively efficient transfer of helium from the garage into the house during the releases, e) the ability of a vehicle to trap a high helium concentration in the engine compartment and, particularly, the undercarriage during a helium release and the relatively rapid drop in these levels to those of the surrounding garage at the end of the release, f) the relatively slow buildup of helium in the passenger compartment and trunk of a vehicle during a helium release and subsequent slow decay following cessation of the flow, g) the effectiveness of active ventilation in reducing helium concentrations in the garage to levels below those corresponding to flammable concentrations of hydrogen, and h) the trapping of helium/air mixtures corresponding to highly flammable hydrogen mixtures inside the vehicles even when active garage ventilation was employed.

Keywords: attached residential garages, automobiles, buoyancy-dominated flow, forced ventilation, Froude number, helium concentration, hydrogen concentration, momentum-dominated flow, natural ventilation, wind effects

Table of Contents

Abstract i
Table of Contents ii
List of Tables iii
List of Figures iv
1. Introduction 1
   1.1 Overview 1
   1.2 Summary of Related NIST Studies 1
   1.3 Related Literature on Hydrogen Dispersion in Residential Garages and Vehicles 2
   1.4 Overview of Gas Exchange Rates in Partially Enclosed Space 4
   1.5 Introduction to the Current Study 6
2   Experimental Set-Up 6
   2.1 Overview and Accident Scenario 6
   2.2 Test House and Garage 8
   2.3 Helium Supply and Release Systems 12
   2.4 Instrumentation 14
      2.4.1 Helium sensors 14
      2.4.2 Pressure measurement 16
      2.4.3 Helium measurements inside the test house 16
      2.4.4 Additional Temperature Measurements 16
   2.5 Experiments with a Vehicle Parked in the Garage 18
   2.6 Experiments with Forced Ventilation 18
3   Experimental Results 19
   3.1 Experiments with Natural Ventilation of the Garage 19
      3.1.1 Helium release from a vertical tube without vehicle 19
      3.1.2 Helium release over large area without vehicle 35
      3.1.3 Helium release over large area with vehicle 44
   3.2 Experiments with Forced Ventilation of the Garage 55
4   Discussion 77
   4.1 Naturally Ventilated Garage Experiments 77
   4.2 Garage Experiments with Forced Ventilation 84
   4.3 Buoyant Gas Trapping within Conventional Automobiles 85
5   Summary 89
6   References 93

List of Tables

Table 1. Helium Sensor Locations for 2008 Tests 15

Table 2. Helium Sensor Locations for 2010 Tests without Vehicle 15

Table 3. Locations of Sensors #9 to #14 inside Vehicle (when present) for 2010 Tests 16

Table 4. Summary of Natural-Ventilation Garage Experimental Flow Parameters 20

Table 5. Helium Concentration Parameters for Garage at the End of Helium Release 24

Table 6. Garage Parameters for Post-Release Period 26

Table 7. Helium Parameters in the House at the End of Helium Release in the Garage 28

Table 8. Summary of Forced Ventilation Garage Experimental Flow Parameters 55

Table 9. Helium Concentration Parameters for Forced-Ventilation Garage Experiments 58

List of Figures

Figure 1 Photograph showing the Indoor Environment and Ventilation Test House and attached garage at NIST. 8

Figure 2 This schematic shows a floor plan for the Indoor Environment and Ventilation Test House, Building 423. 8

Figure 3 This schematic shows a floor plan for the garage attached to the Indoor Environment and Ventilation Test House, Building 423. 11

Figure 4 An interior view of the garage looking toward the front in which the floor and garage door can be seen. 12

Figure 5. An interior view of the garage looking toward the rear is shown. The exterior doorway on the rear wall can be seen. The helium release flow conditioner used for the 2010 experiments is visible on the floor in the foreground. 13

Figure 6. Photograph showing the test vehicle parked inside the garage. 17

Figure 7. Photograph showing the OSB board with attached flange in place in the rear doorway of the garage. The flange was connected to the Infiltec Model DL1-DM4-110 duct leakage tester (visible at the base of the doorway) by flexible duct. 19

Figure 8 Measured helium volume flow rate is plotted as a function of time for the 9/11/08 garage experiment. 20

Figure 9. Helium volume percent is plotted as a function time for the eight sensor heights indicated for the 9/11/08 experiment. 21

Figure 10. Helium volume percent is plotted as a function time for the eight sensor heights indicated for the initial 1000 s of the 9/11/08 experiment. 21

Figure 11. Helium volume percent is plotted as a function time for the eight sensor heights indicated over the period from 6300 s to just after the end of the helium release for the 9/11/08 experiment. The color codes for the sensor locations are the same as in Figure 10. 22

Figure 12. Time variations of wind velocity and the differential pressure between the garage interior and outside ambient are plotted over the time period from 6300 s to 14400 s for the 9/11/08 experiment. 23

Figure 13. Experimental helium volume percents measured along the vertical array at the end of the 9/11/08 release are shown along with extrapolated values at the floor and ceiling. The solid lines are approximations of the data that are used to estimate the average helium concentration along the vertical direction. 24

Figure 14. Helium volume percent is plotted as a function time at the eight sensor heights indicated for the period from 43 200 s to 90 000 s in the 9/11/08 experiment. 25

Figure 15. Helium volume percent recorded by Sensor #4 is plotted as a function time for the 9/11/08 experiment. The red line is the result of fitting the data to an exponential decay using a nonlinear least squares curve fit. 26

Figure 16 Temperatures measured using thermocouples (symbols) and thermistors (lines) are plotted as a function of time for measurements recorded on 9/11/08 in the family room, garage, and outside. 27

Figure 17. The helium volume percent measured in the family room of the house is shown as a function of time for the 9/11/08 experiment. 28

Figure 18. Helium volume percent is plotted as a function of time at the eight sensor heights Indicated for the 9/12/08 experiment. 29

Figure 19. Time variations of wind velocity and the differential pressure between the garage interior and outside ambience are plotted over the time period from 6300 s to 14400 s for the 9/12/08 experiment. 30

Figure 20. Wind speeds and differential pressure measurements are plotted over the twelve hours following the start of the experiments on 9/11/08 and 9/12/08. 32

Figure 21. Wind speed and direction are plotted as a function of time for the 9/11/08 and 9/12/08 experiments. 33

Figure 22. Temperatures measured using thermocouples (symbols) and thermistors (lines) are plotted as a function of time for measurements recorded on 9/12/08 in the family room, garage, and outside. 34

Figure 23. The helium volume percent measured in the family room of the house is shown as a function of time for the 9/12/08 experiment. 34

Figure 24. Helium volume percent is plotted as a function time at the nine sensor heights indicated for the 7/29/10 experiment. 35

Figure 25. Experimental helium volume percents measured along the vertical array at the end of the 7/29/10 release are shown along with extrapolated values at the floor and ceiling. The solid lines are approximations of the data that are used to estimate the average helium concentration along the vertical direction. 36

Figure 26. Helium volume percent is plotted as a function of time at the indicated heights for four sensors above the helium release location (center) and a single sensor at (x,y) = (4.343 m, 1.816 m) (front-right) for the 7/29/10 experiment. 37

Figure 27. Helium volume percent is plotted as a function of time for the five sensor heights indicated for the initial 300 s of the 7/29/10 experiment. 38

Figure 28. Temperatures measured using thermocouples (symbols) and thermistors (lines) are plotted as a function of time for measurements recorded on 7/29/10-7/30/10 in the family room, garage, and outside. 39

Figure 29. The helium volume percent measured in the family room of the house manually (symbol) and electronically (line) is shown as a function of time for the 7/29/10 experiment..............40

Figure 30. Helium volume percent is plotted as a function of time at the nine sensor heights indicated for the 8/2/10 experiment...40

Figure 31. Helium volume percent is plotted as a function time at the indicated heights for four sensors above the helium release location (center) and a single sensor at (x,y) = (4.343 m, 1.816 m) (front-right) for the 8/2/10 experiment..41

Figure 32. Helium volume percent is plotted as a function time for the five sensor heights indicated for the initial 300 s of the 8/2/10 experiment..42

Figure 33. The time variation of the differential pressure between the garage interior and outside is plotted over the time period from 0 s to 43 200 s for the 8/2/10 experiment. 42

Figure 34. Temperatures measured using thermocouples (symbols) and thermistors (lines) are plotted as a function of time for measurements recorded on 8/2/10 in the family room, garage, and outside. 43

Figure 35. The helium volume percents measured in the family room of the house manually (symbols) and electronically (line) are shown as a function of time for the 8/2/10 experiment. 43

Figure 36. Helium volume percent in the garage is plotted as a function time at the eight sensor heights indicated along the vertical array for the 8/6/10 experiment.....................................44

Figure 37. Experimental helium volume percents measured along the vertical array at the end of the 7/29/10 release are shown along with extrapolated values at the floor and ceiling. The solid lines are approximations of the data used to estimate the average helium concentration along the vertical direction...45

Figure 38. The time variation of the differential pressure between the garage interior and outside is plotted over the time period from 0 s to 43 200 s for the 8/6/10 experiment. 47

Figure 39. Temperatures measured using thermocouples (symbols) and thermistors (lines) are plotted as a function of time for measurements recorded on 8/6/10 in the family room, garage, and outside. 48

Figure 40. The helium volume percents measured in the family room of the house manually (symbol) and electronically (line) are shown as a function of time for the 8/6/10 experiment. 48

Figure 50. Experimental helium volume percents (solid symbols) measured along the vertical array at the end of the 8/9/10 release are shown along with extrapolated values (open symbols) at the floor and ceiling. The solid lines are approximations of the data used to estimate the average helium concentration along the vertical direction. 57

Figure 51. Helium volume percent recorded by Sensor #4 is plotted as a function of time for the 8/9/10 experiment. The red line is the result of fitting the data to an exponential decay using a nonlinear least squares curve fit. 58

Figure 52. Helium volume percent is plotted as a function of time at the indicated locations in the automobile centered over the helium release location for the 8/9/10 experiment with forced ventilation. 59

Figure 53. Temperatures measured using thermocouples (symbols) and thermistors (lines) are plotted as a function of time for measurements recorded on 8/9/10 in the family room, garage, and outside. 60

Figure 54. The helium volume percent in the family room of the house measured electronically is shown as a function of time for the 8/9/10 experiment. 60

Figure 55. Helium volume percent is plotted as a function of time for the eight sensor heights indicated for the 8/12/10 experiment with forced ventilation. 61

Figure 56. Helium volume percent is plotted as a function of time at the indicated locations in the automobile centered over the helium release location for the 8/12/10 experiment with forced ventilation. 63

Figure 57. Temperatures measured using thermocouples (symbols) and thermistors (lines) are plotted as a function of time for measurements recorded on 8/12/10 in the family room, garage, and outside. 64

Figure 58. The helium volume percent measured in the family room of the house electronically is shown as a function of time for the 8/9/10 experiment. 64

Figure 59. Helium volume percent is plotted as a function of time for the eight sensor heights indicated for the 8/13/10 experiment with forced ventilation. 65

Figure 60. Experimental helium volume percents (solid symbols) measured along the vertical array at the end of the 8/13/10 release are shown along with extrapolated values (open symbols) at the floor and ceiling. The solid lines are approximations of the data used to estimate the average helium concentration along the vertical direction. 66

Figure 61. Temperatures measured using thermocouples (symbols) and thermistors (lines) are plotted as a function of time for measurements recorded on 8/13/10 in the family room, garage, and outside. 66

Figure 62. Helium volume percent is plotted as a function time at the indicated locations in the automobile centered over the helium release location for the 8/13/10 experiment with forced ventilation. 67

Figure 63. Helium volume percent is plotted as a function time for the eight sensor heights indicated for the 8/13/10a experiment with forced ventilation. 68

Figure 64. Experimental helium volume percents (solid symbols) measured along the vertical array at the end of the 8/13/10a release are shown along with extrapolated values (open symbols) at the floor and ceiling. The solid lines are approximations of the data used to estimate the average helium concentration along the vertical direction. 69

Figure 65. Temperatures measured using thermocouples (symbols) and thermistors (lines) are plotted as a function of time for measurements recorded on 8/13/10a in the family room, garage, and outside. 69

Figure 66. Helium volume percent is plotted as a function of time at the indicated locations in the automobile centered over the helium release location for the 8/13/10a experiment with forced ventilation. 70

Figure 67. Helium volume percent is plotted as a function of time for the eight sensor heights indicated for the 8/19/10 experiment with forced ventilation. 71

Figure 68. Experimental helium volume percents (solid symbols) measured along the vertical array at the end of the 8/19/10 release are shown along with extrapolated values (open symbols) at the floor and ceiling. The solid lines are approximations of the data used to estimate the average helium concentration along the vertical direction. 73

Figure 69. Temperatures measured using thermocouples (symbols) and thermistors (lines) are plotted as a function of time for measurements recorded on 8/19/10 in the family room, garage, and outside. 73

Figure 70. Helium volume percent is plotted as a function of time at the indicated locations in the Passat centered over the helium release location for the 8/19/10 experiment with forced ventilation. 74

Figure 71. Helium volume percent is plotted as a function time for the eight sensor heights indicated for the 8/19/10a experiment with forced ventilation. 75

Figure 72. Experimental helium volume percents (solid symbols) measured along the vertical array at the end of the 8/19/10a release are shown along with extrapolated values (open symbols) at the floor and ceiling. The solid lines are approximations of the data used to estimate the average helium concentration along the vertical direction. 75

Figure 73. Temperatures measured using thermocouples (symbols) and thermistors (lines) are plotted as a function of time for measurements recorded on 8/19/10a in the family room, garage, and outside. 76

Figure 74. Helium volume percent is plotted as a function of time at the indicated locations in the Passat centered over the helium release location for the 8/19/10a experiment with forced ventilation. 77

Figure 75. Helium volume percent is plotted as a function of time for sensors located 2.134 m above the floor immediately above the helium release location and in the front-right quadrant of the garage for the 7/29/10 experiment. 78

Figure 76. The falloff of helium volume percent in the family room of the house following the end of the helium release inside the garage on 7/29/10 is fit to an exponential curve using a least squares curve fitting procedure. 83

Figure 77. The falloff of helium volume percent in the passenger compartment of the vehicle long after the end of the helium release under the vehicle on 8/6/10 is fit to an exponential curve using a least squares curve fitting procedure. 88

viii

1. Introduction

1.1 Overview

Concerns about the potential effects of the buildup of high carbon dioxide concentrations in the earth's atmosphere due to burning of hydrocarbon fuels have led to efforts to develop technologies that utilize hydrogen as an energy carrier. One active area has been the commercial development of hydrogen-fueled fuel cell vehicles, with initial deployment currently predicted in less than five years. Due to substantial differences in physical properties and burning behavior between hydrogen and hydrocarbon fuels, a great deal of effort has been placed on the fire safety implications of the wide-spread use of hydrogen. One major focus has been the potential for hydrogen to be accidentally released into partially enclosed spaces such as fuel cell enclosures, buildings, and commercial and residential garages.

Both cryogenic liquid and compressed hydrogen stored in appropriate tanks have been utilized in prototype hydrogen-fueled vehicles. In the short term, it appears that high-pressure gaseous storage will be dominant. Thus the potential for accidental releases of hydrogen gas should be considered. Under the assumption that a large percentage of a fleet of hydrogen-fueled vehicles will be housed in the existing stock of residential garages, the National Institute of Standards and Technology (NIST) has undertaken an effort to characterize the hydrogen distribution and potential combustion behavior during and following the accidental release of hydrogen into a typical residential garage.

1.2 Summary of Related NIST Studies

An initial experimental effort utilizing helium as a surrogate for hydrogen investigated the effects of helium release rate, release location, and vent size and location on the distribution of and leakag of helium from an idealized ¼-scale two-car garage. A NIST Technical Note and a manuscript presented during the Third International Conference on Hydrogen Safety describe this work. [1,2] Detailed time records of the measurements have been posted on the World Wide Web to provide well-characterized data sets for validating and improving models of the mixing process. [3] The total helium volume released was scaled to be equivalent to the release of 5 kg of hydrogen into the full-scale garage. Vent sizes were scaled to provide the equivalent of three garage air changes per hour, $(ACH_{gar})_{4Pa}$, for a pressure difference of 4 Pa between the interior and surroundings of the garage. One and four hour releases at the lower center, lower rear, and upper center of the garage were studied. These experiments were used to characterize the ability of a computational code to predict such flows. [4]

A follow-on study investigated the effect of replacing the poly(methyl methacrylate) front wall of the ¼-scale garage with gypsum drywall. It was shown that helium readily diffuses through the dry wall, and the helium concentration drops rapidly inside following the end of the helium release. [5] The diffusion of helium through the dry wall is so much faster than the reverse diffusion of air, that the pressure inside the garage dropped rapidly by hundreds of Pascals when the garage was sealed. Experiments in which forming gas containing low levels of hydrogen (≈ 4 %) in nitrogen were released into the garage showed that the diffusion rate for hydrogen through drywall was similar to that for helium.

NIST sponsored a study carried out by Southwest Research Institute (SWRI) in which hydrogen was released into a structure having the dimensions of a two-car garage. Doorway fan tests were used to characterize the global leak characteristics of the structure before each test. Hydrogen was released at low speed near the center of the floor. The constant hydrogen volume flow rate was sufficient to release 5 kg in one hour. Experiments were performed with an empty garage and with a variety of vehicles centered inside. A range of instrumentation, including thermal conductivity sensors, thermocouples, piezoelectric pressure transducers, and different types of video cameras, was used to characterize the hydrogen volume fraction distribution and time behavior as well as flame spread rates and overpressures when the hydrogen was ignited. Ignition was initiated when the hydrogen level reached one of a number of pre-determined values. Post-analysis of the raw experimental data was performed by both SWRI and

1

NIST researchers. Details concerning the experiment are available in a NIST Government Contractor Report [6], and a manuscript describing the experimental findings is available [7].

Two related studies were carried out in conjunction with work described in this report. The first was a modeling study of the effectiveness of forced ventilation in preventing the buildup of dangerous concentrations of hydrogen within a full-scale garage attached to a house, both with and without a vehicle present. [8] The second investigated the response of hydrogen detectors placed in the garage to hydrogen released as forming gas. [9]

A number of computational fluid mechanic and analytical studies by NIST have considered hydrogen release in a garage subject to a range of conditions. [10,11,12]

1.3 Related Literature on Hydrogen Dispersion in Residential Garages and Vehicles

The prospect of the deployment of a fleet of hydrogen fuel cell powered automobiles has raised concerns about the potential hazards associated with releases of hydrogen in partially enclosed spaces such as commercial and residential garages. As a result, a large number of experimental and modeling studies describing the distribution of hydrogen (or helium serving as a surrogate) have been reported. We have recently summarized such studies in a report describing the mixing of helium released within an idealized ¼-scale two-car garage, which can be consulted for a more complete review. [1] Here we focus on studies directly related to the current study on mixing in residential garages and within vehicles.

Cariteau et al. considered the distribution of helium generated by various types of helium releases at different locations under a vehicle and in the empty garage. [13] The enclosure was scaled as a single-car garage, 5.76 m long × 2.96 m wide × 2.42 m high, and was very well sealed with a measured air change per hour rate, ACH_{gar} of 0.01 h^{-1}. During each experiment, a total helium volume of 1.09 m^3 was released at a constant rate for either 105 s or 316 s. Upward helium releases from a 7 cm diameter source in the empty garage showed that the higher flow rate resulted in higher concentrations near the ceiling at the end of a release. Vertical concentration profiles showed that the helium distribution was partially stratified, with the higher flow rate resulting in a more stratified upper layer. Releases at two different horizontal locations in the garage had little effect on the helium distribution.

Helium was also released at several locations under a small commercial van either as a diffuse upward source located under the engine compartment hood or beneath the engine or as a downward jet located beneath the engine or passenger compartment. [13] The varying release conditions resulted in different helium concentrations near the ceiling as well as changing the vertical concentration distribution. The helium distributions for the releases under the hood were very similar to those observed when the vehicle was not present. When the diffuse source was moved below the engine, the helium concentration near the ceiling fell slightly, and the vertical distribution was not as stratified. The downward releases gave very different results. The helium concentrations near the ceiling were significantly decreased and the vertical helium concentration profile was nearly uniform. Cariteau et al. attributed these observations to changes in the mixing with the effective areas of the release. The downward jets effectively created large-area releases that increased mixing and limited stratification, while smaller area releases encouraged stratification. [13]

Maeda et al. performed a study of hydrogen buildup inside the front engine compartment of a conventional rear-drive passenger sedan. [14] Hydrogen was released at one of two locations near the ground—centered between the front wheels and centered under the passenger compartment. Various combinations of hydrogen volume flow rates (minimum of 9×10^{-5} m^3/s to a maximum of 2.31×10^{-3} m^3/s, assuming a release temperature of 20 °C) and release periods (varying from 30 s to 600 s) were used. Hydrogen concentrations were measured in the engine compartment near the center just underneath the hood, at the front at the top of the radiator, and at the front near the bottom of the radiator. Quasi-steady concentrations of hydrogen developed quickly during the releases, with levels depending on release point, measurement location, and flow conditions. For a given flow condition, concentrations were higher for release under the passenger compartment as compared to under the engine.

2

Measured quasi-steady-state concentrations were generally higher at the measurement location under the hood, with slightly lower concentrations measured at the top of the radiator. Hydrogen concentrations at the bottom of the radiator were typically close to zero. The highest measured hydrogen volume percents of approximately 24 % were observed for the higher hydrogen volume flow rates and longer release periods. Once the hydrogen flow was halted, concentrations within the engine compartment dissipated quickly, requiring less than 200 s to fall below the lower flammability limit concentration for hydrogen.

Flammable mixtures of hydrogen and air (> 4 % hydrogen) were observed during these experiments. Spark ignition of these mixtures was used, and the resulting flames were characterized by temperature, thermal radiation, and sound intensity measurements. These authors concluded that the conditions generated outside of the engine compartment were not severe enough to cause major injuries or fatalities to people near the vehicle. [14]

Maeda et al. extended the findings of their original study [14] of mixing in the engine compartment of a vehicle to higher hydrogen volume flow rates, while adding an additional release location and several hydrogen sensor locations. [15] Hydrogen volume flow rates for 600 s releases were varied over a range from 3.52×10^{-3} m^3/s to 0.0176 m^3/s. Hydrogen sensors were added at the ventilator louver between the hood and windshield and at two vertical locations along a line between the bottom side of the hood and the top of the engine. The additional release location was centered under the rear axle of the rear-drive vehicle. Hydrogen was released either upwards or downwards from a 4 mm diameter tube.

For upward releases, hydrogen concentrations with volume flow rates above 3.53×10^{-3} m^3/s were nearly uniform throughout the upper volume of the engine compartment. In the lower volume of the engine compartment (bottom of the front bumper), no hydrogen was measured. These observations suggest that sharp concentration gradients were present at intermediate heights in the engine compartment. For lower volume flow rates, some stratification in the upper part of the engine compartment was evident. The quasi-steady-state concentrations increased relatively slowly with increasing volume flow rate. [15]

Downward releases under the vehicle resulted in changes to the distribution of hydrogen within interior volumes. For a given hydrogen volume flow rate, the quasi-steady-state concentration was lower than for the corresponding upward release, and there was a stronger dependence on the volume flow rate. Stratification of the hydrogen concentration in the upper volume of the engine compartment was observed over the entire volume flow rate range. Unlike for upward releases, measurable hydrogen concentrations were present at the lower measurement location at the front of the engine compartment.

By covering the drive shaft tunnel under the vehicle with a flat metal sheet, the authors demonstrated that the geometry present under the vehicle had a large effect on the hydrogen concentrations trapped in the engine compartment. A similar dependence on release location was identified, with the highest concentration observed when the release point was centered under the passenger compartment.

Since hydrogen concentrations were higher in the second round of experiments, more vigorous combustion was observed when the hydrogen/air mixtures inside the engine compartment were ignited at the end of the release period. For the highest levels some deformation of the hood of the vehicle was reported. Even so, the authors reached a similar conclusion to their earlier work that such an ignition would be unlikely to cause major injury to a person located nearby, but outside the vehicle. [15]

Merilo et al. investigated the mixing and combustion behavior of hydrogen when released into a hardened building equipped with a plastic-sheet front wall and scaled to represent a single-car garage (3.64 m (w) × 6.10 m (l) × 2.72 m (h)). [16] Hydrogen was released from a 7.75 mm (ID) diameter tube at volume flow rates (assuming a 20 °C release temperature) between 2.9×10^{-3} m^3/s and 0.030 m^3/s. The garage was equipped with fan ventilation, and the majority of the experiments were cases where active ventilation and hydrogen releases rates were varied for hydrogen flows into the empty garage in order to assess the effectiveness of the ventilation for limiting the buildup of hydrogen and the potential for burning.

Three experiments were run using natural ventilation with vents designed to meet the recommendations of the 2002 ICC Final Action Agenda, which included recommendations for residential

3

garages. [17] These recommendations referred to cases where hydrogen generation or refueling operations were present and specifically excluded parked hydrogen-fueled vehicles in garages. The requirements were based on experiments and modeling in a real-scale single car garage containing a vehicle mock up carried out at the University of Miami. [17] Hydrogen scenarios considered were 20 min releases at 5.1×10^{-4} m^3/s, 2.0×10^{-3} m^3/s, and 6.1×10^{-4} m^3/s. The vents were required to be located near the floor and ceiling and to have minimum areas of 0.0465 m^2 per 28.3 m^3 of garage volume.

Two of the experiments with natural ventilation were run for hydrogen volume flow rates near 0.030 m^3/s (assuming a temperature of 20 °C). The first was done with an empty garage and the second with a vehicle present. For the latter, the hydrogen release location was outside of the vehicle near the expected location of a refueling port. The third experiment was run at a lower volume flow rate of 2.9×10^{-3} m^3/s with an empty garage. Hydrogen concentrations were monitored every 5 min to 10 min at three heights within the upper 1/3 of the garage. The measurements showed that the upper layer of the garage was well mixed for experiments where higher hydrogen volume flow rates were used. This was attributed to the mixing induced by overturning when the momentum-dominated flow reached the ceiling. For the lower flow rate case, some stratification of the upper layer was evident, which was judged to be consistent with a buoyancy-dominated flow. Maximum hydrogen volume percents at the ends of the releases near the ceiling were around 22 % for the higher volume flow rates and between 2.9 % and 7.1 % for the lower volume flow rate.

Ignition of the higher concentration levels resulted in observed maximum flame spread rates as high as 60 m/s. Ignition of the mixture at the lower release rate was not described. Overpressure measurements were reported for the higher concentration cases with and without a vehicle. Measurable overpressures were present for both experiments, but the magnitudes and fluctuations were much larger when the vehicle was present. The higher pressure was attributed to an internal explosion within the vehicle. The observed damage of the vehicle was consistent with this conclusion and showed that burning occurred in the passenger compartment as well as the engine compartment. [16]

Ekoto et al. investigated the interior hydrogen concentration distributions generated by releasing hydrogen or helium directly into the passenger compartment or trunk of a conventional automobile. [18] Releases were done in such a way that they were either momentum dominated or buoyancy dominated. Three volume flow rates of released gas, 8.3×10^{-4} m^3/s, 4.2×10^{-4} m^3/s, and 1.97×10^{-3} m^3/s, were used. Major findings were that 1) flammable concentrations of hydrogen developed in the release compartment over short periods of time (order of seconds), 2) very high levels of hydrogen were observed at longer times for the higher flow rates, 3) quasi-steady states developed over periods of tens of minutes which depended on release rates and locations, 4) momentum-dominated releases resulted in more homogeneous mixing, 5) releases in the passenger compartment led to high concentrations in the trunk, but the reverse was not the case, and 6) for the same volume flow rates helium concentrations were measurably lower than recorded with hydrogen.

Liu and Schreiber employed a computational fluid dynamics (CFD) model to study the similar problem of the release of hydrogen into a passenger compartment of a conventional automobile with active ventilation. [19] The hydrogen volume flow rate was assumed to be 3.1×10^{-3} m^3/s, and the compartment was actively ventilated with air at an unspecified rate. With the base configuration, the average hydrogen helium volume percent when a pseudo-steady state developed was 4.6 %, with over 60 % of the compartment above the 4.1 % hydrogen flammability limit. The authors showed that moving the exhaust vents for the passenger compartment to the rear ceiling significantly reduced the passenger compartment hydrogen concentrations, with only 3.1 % of the volume having values greater than the lower flammability limit.

1.4 Overview of Gas Exchange Rates in Partially Enclosed Space

Leak rates for a partially enclosed space with a volume, V_{enc}, are typically characterized in terms of the number of air changes per hour (ACH_{enc}), which corresponds to a volume flow exchange rate across the enclosure boundary, Q_{enc}, given by $Q_{enc} = V_{enc} \times ACH_{enc}/3600$ s/h. ACH_{enc} can vary substantially with time and depends not only on the areas and characteristics of openings connecting across the enclosure boundary, but also on such factors as weather conditions and forced ventilation. Values of Q_{enc} can be related to an effective leak area, ELA_{enc} by use of the Bernoulli equation,

$$(ELA_{enc})_H = \frac{(Q_{enc})_H}{\left(2\frac{\Delta P}{\rho}\right)^{\frac{1}{2}}}$$

(1)

where the subscript H indicates evaluation at a specific pressure difference, ΔP, between the interior and exterior and gas density, ρ. A value of $H = 4$ Pa is often taken to be representative of pressure differences across a garage boundary that develop due to wind and temperature differences between the interior and ambience. [20] A table (their Table 1) included in a report by Chan et al. suggests that 4 Pa is on the high side, and actual values typically range from 0.2 Pa to 2 Pa for "mild" to "severe" weather conditions. [21] Note that $(ELA_{enc})_H$ does not normally correspond to the actual open area in a garage boundary since experimental flow rates typically obey the following relation [20]

$$Q_{enc} = C \times \Delta P^n,$$

(2)

where C is the flow parameter and n is an experimental parameter that varies between 0.5 and 1. [20]

Studies indicate that values of $(ACH_{gar})_{4Pa}$ and $(ELA_{gar})_{4Pa}$ vary widely for residential garages in the United States, e.g., see [20,22]. It is reasonable to consider values recommended as the minimum acceptable required levels. A recommended minimum value of $Q_{gar} = 2.73$ m^3/min (100 ft^3/min) per stored automobile was included in a version of an American Society of Heating, Refrigeration and Air Conditioning Engineers (ASHRAE) standard. [23] Swain et al. referred to this value in an earlier study of hydrogen release in garages. [24] Note that this recommendation is no longer included in current versions of the ASHRAE standard, but is incorporated in the 2009 International Mechanical Code published by the International Code Council (ICC). [25] A simple calculation reveals that this value corresponds exactly to an $ACH_{gar} = 3$ h^{-1} for a single car garage sized 3.048 m (w) × 6.096 m (l) × 3.048 m (h). On this basis, an $(ACH_{gar})_{4\,Pa} = 3$ h^{-1} was adopted as the recommended lower value for our reduced-scale study. [1,2]

Values of $(ACH_{enc})_{4Pa}$ are usually determined with a device known as a blower door that measures the ΔP across a boundary for a given volume flow rate of air, Q_{enc}, into or from the space. Note that measurements are frequently taken at several higher pressures and extrapolated to 4 Pa using Eq. (2). Results described in [26] for 67 attached garages in four Canadian cities are reported in terms of $(ACH_{gar})_{50\,Pa}$. Averages can be roughly converted to $H = 4$ Pa values using Eq. (2) and an assumed value of $n = 0.65$. The results are $(ACH_{gar})_{4\,Pa} = 7.2$ h^{-1}, 3.5 h^{-1}, 3.3 h^{-1}, and 9.1 h^{-1} for Vancouver, Winnipeg, Saskatoon, and Ottawa, respectively. Emmerich et al. reported findings that can be converted to $(ACH_{gar})_{4\,Pa}$ for five residential garages in the Washington, DC area. Values ranged from $(ACH_{gar})_{4\,Pa} = 1.4$ h^{-1} to 25.3 h^{-1}. [22]

An approach known as the tracer gas method allows measurement of ACH_{enc} values by either releasing short bursts or a continuous stream of a tracer gas into the enclosure and tracking its concentration as a function of time. Note that such values are expected and are observed to vary with time as conditions change. Several studies have reported such values for residential garages. Emmerich et al. summarized some early studies. [22] Studies of garages in the United States and Canada yielded ACH_{gar} values covering a range from 0.3 h^{-1} to 2.7 h^{-1}. The most common values were between 0.7 h^{-1} and 1.0 h^{-1}. More recently, Batterman et al. reported ACH_{gar} measurements of fifteen residential garages in Michigan. [27] ACH_{gar} values ranged from 0.16 h^{-1} to 1.80 h^{-1}, with an average of 0.77 h^{-1} and standard deviation of 0.51 h^{-1}. Waterland et al. measured time-varying values of ACH_{gar} for three garages in Texas and California. [28] Average values of ACH_{gar} were approximately 0.5 h^{-1}, 0.5 h^{-1}, and 0.2 h^{-1}. Together, these findings suggest that typical residential garages in North America have ACH_{gar} values for ordinary weather conditions between 0.5 h^{-1} and 1 h^{-1}, with a most probable value near 0.7 h^{-1}. Note that most

5

measured ACH_{gar} values are more than a factor of three lower than the minimum level recommended by ASHRAE and the ICC. [23,25] Some of the studies above were reviewed by Adams et al. who provided a probability plot for ACH_{gar} values that is consistent with, but skewed to somewhat lower values, than suggested above. [29] As a worst case value, they recommended a minimum value of $ACH_{gar} = 0.03$ h^{-1}, i.e., 100 time less than the recommended minimum value, which appears to be very conservative.

1.5 Introduction to the Current Study

As discussed above and in a NIST technical note [1], there has been a great deal of recent activity on understanding the mixing behavior of hydrogen (or helium acting as a surrogate) when released into partially enclosed spaces. Much of this interest is based on concern about potential accidental releases from hydrogen fueled automobiles parked in residential garages. Concentration distributions have been measured in enclosures representing residential garages as well as within conventional automobiles.

Most of the experiments have been performed in specially constructed structures designed to represent residential garages. Potential leaks to the outside have generally been minimized or controlled by creating idealized vents. Most experiments have been designed to minimize temperature and wind variations, even though such effects are known to dominate air exchanges between structures and their surroundings. To our knowledge, no experiments have been performed with a garage attached to a residence, even though a large fraction of residential garages in the United States are attached.

In the study reported here, twelve experiments were performed in which helium, acting as a surrogate for hydrogen, was released into a garage attached to a house. Six of these tests involved natural ventilation of the garage, and six used a fan to create forced ventilation. For the natural ventilation cases, helium was released as either momentum- or buoyancy-dominated flow. Experiments were performed with and without a conventional vehicle parked inside the garage. A vehicle was present during all six of the forced-ventilation experiments. The distribution of helium within the garage was monitored by time-resolved measurements at multiple locations and for a single location inside the attached house. When a vehicle was parked inside, helium concentrations were measured below the undercarriage, in a wheel well, at two locations inside the engine compartment, in the passenger compartment, and in the trunk. Supplemental measurements included wind speed and direction, differential pressures, and temperatures at various locations.

2 Experimental Set-Up

2.1 Overview and Accident Scenario

The experiments were designed to study the mixing behavior and concentration buildup when hydrogen is accidentally released from a hydrogen-fueled automobile parked in a residential garage attached to a single family home. Due to safety concerns, helium was used as a surrogate gas for hydrogen. As discussed above and in reference [1], numerous studies have demonstrated that the behaviors of hydrogen and helium are similar in experiments designed to study mixing and dispersion in partially enclosed spaces.

The highly idealized accident scenario considered was the release of 5 kg of pressurized hydrogen gas (a typical amount for full tanks on existing models of hydrogen-fueled automobiles) at a constant rate over a period of 4 h. Assuming an ideal gas at a temperature of 20 °C, this corresponds to a volume of 59.7 m^3. The flow rate required to release this volume over 4 h is 4.15 L/s. This volume flow rate was chosen as the target value for the helium experiments.

Helium was released near the floor at the center of the garage. Cases were studied in which the garage was empty as well as when an automobile was parked directly over the release location. The experiments were designed to characterize the mixing and distribution of helium within the garage, the loss of the helium from the garage, and the buildup and decay of helium inside the house both during the release and in the post-release period. When the automobile was present, helium concentrations were

Figure 1. Photograph showing the Indoor Environment and Ventilation Test House and attached garage at NIST.

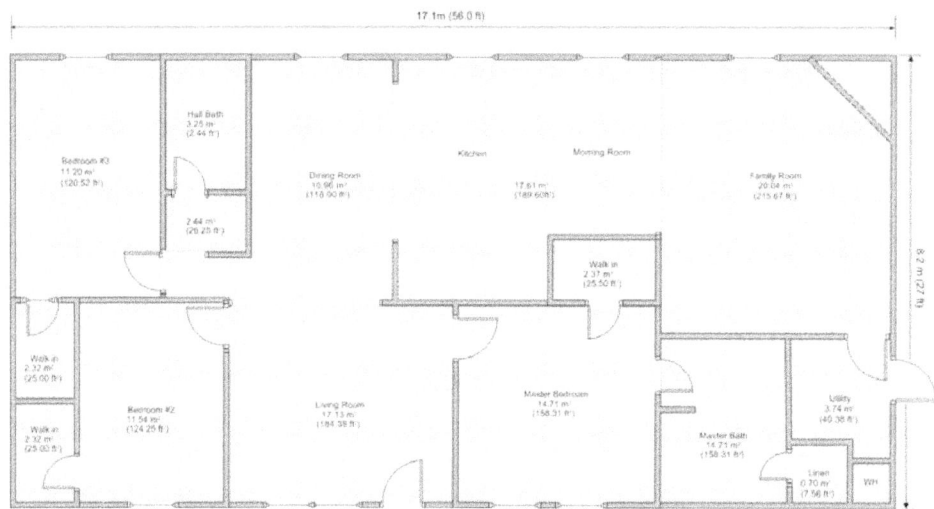

Figure 2. This schematic shows a floor plan for the Indoor Environment and Ventilation Test House, Building 423.

monitored at multiple interior locations in order to characterize the potential buildup of hydrogen concentrations in such spaces.

Two sets of experiments were performed during September of 2008 and late July and early August of 2010.

2.2   Test House and Garage

The single-car garage employed for this study is attached to Building 423 at NIST, which is known as the "Indoor Environment and Ventilation Test House." An exterior photograph of the house and garage from the front is shown in Figure 1. Detailed descriptions of the house are available in two NIST Reports [30,31] which describe the characteristics of the house as originally built and following an airtightening refit. The single-story manufactured home has an outside width of 8.2 m and length of 17.1 m. It is oriented with its front facing towards the north. The house has three-bedrooms and two-baths with a large open area consisting of a living room, dining room, kitchen, morning room and family room. Figure 2 shows the floor plan. The house is equipped with a pitched roof above a low attic space separating the

8

roof and an interior cathedral ceiling that runs along the full length of the house. The ceiling is 2.7 m above the floor at the center and slopes down to a height of 2.1 m at the front and rear walls. The internal floor area and volume are 140 m^2 and 340 m^3, respectively.

The house is heavily instrumented with sensors at multiple locations to measure such properties as interior and exterior temperatures, differential pressure between the interior and exterior, atmospheric pressure, relative humidity, fan operation, duct flow rates, and energy usage. Tracer gas measurements using sulfur hexafluoride were used to characterize air changes per hour for the house, ACH_{hou}. [30,31] Ceramic-coated thermistors were used to measure temperatures at multiple locations within the house as well as the outdoor temperature on a weather tower. A sonic anemometer placed at the top of a 10 m high tower was used to record wind speed and direction at a location 5 m to the south of the house. The measurements are recorded continuously using a dedicated data acquisition system at a rate of 1 min^{-1}. Additional details concerning the sensor locations and uncertainties are available in the reports. [30,31]

Measurements showed that ACH_{hou} values decreased marginally following the airtightness refit. As expected, ACH_{hou} values were found to increase with increasing wind speed and temperature difference between the interior and outside ambience. Observed values varied over a range from roughly 0.1 h^{-1} to 0.4 h^{-1} following the refit. [31] The range was somewhat wider, with values up 0.8 h^{-1}, for the as-built house. [30,31]

The weather measurement system described above was functional for the 2008 experiments. During the second series of experiments the anemometer malfunctioned, and local wind speed data is unavailable.

The attached garage was constructed on the west end of the structure after the house had already been updated for airtightness. A schematic floor plan for the garage is shown in Figure 3. It has a concrete foundation and poured concrete floor. The garage walls were constructed using standard "2×4" wood stud construction on 0.41 m centers. Outside walls were covered with plywood sheathing and exterior siding. The original siding between the house and the garage was removed and replaced with dry wall. The roof has a central height of 3.66 m and the same slope as the roof of the house.

The interior dimensions of the garage are 5.33 m (w) × 6.71 m (l) × 2.43 m (h) with a 36.5 m^2 floor area and 86.9 m^3 internal volume. The walls and ceiling are covered with finished, primed, and painted gypsum drywall with painted wooden molding around the lower edge. The drawings for the garage specify the use of 1.27 cm thick drywall and a one hour fire rating for the wall assembly. This implies the use of Type X fire-rated drywall, and inspection indicated that Type X had been used.

The north wall has a four-section 2.74 m (w) × 2.13 m (h) roll-up garage door equipped with outside sealing gaskets along the four edges. The side and top gaskets are sealed against 0.64 cm thick molding attached to the door frame, while the bottom gasket is press sealed against the concrete base. The effective open area of the doorway is 2.64 m (w) × 2.08 m (h). It appeared to be well sealed, and there were no obvious leaks visible when closed. The garage door is shown from the interior in Figure 4.

There are two additional doors into the garage. One is an exterior entrance with dimensions of 0.914 m (w) × 2.032 m (h) located on the left side of the rear wall. This door is sealed against the exterior door frame molding on the sides and top and at the bottom by a gasket. This door can be seen in Figure 5. Inspection revealed ≈ 0.3 cm gaps around the door through which light was visible coming from the outside with the door closed. The second door connects to the house and is located where an original side entrance entered the manufactured home (see Figure 2). It has the same dimensions as the rear door. This door was well sealed at the bottom by a gasket pressed against floor molding. Elsewhere, there were gaps of ≈ 0.3 cm between the door and frame, including between the strike legs on the sides and the header at the top. Light was visible passing between the garage and interior at these locations. The doorway passed into a small utility room (see Figure 2), which was, in turn, connected to the family room by a second door. There was a 10 cm gap between the bottom of this second door and the floor.

The garage was constructed with two 0.902 m (w) × 1.219 m (h) windows located on the right rear wall and on the back side of the exterior side wall. The side-wall window had been removed, and the opening was sealed with a plywood sheet holding a centered 0.65 m × 0.65 m louvered aluminum box containing an exhaust fan. For these experiments, the fan was not operated, and the louvers were sealed.

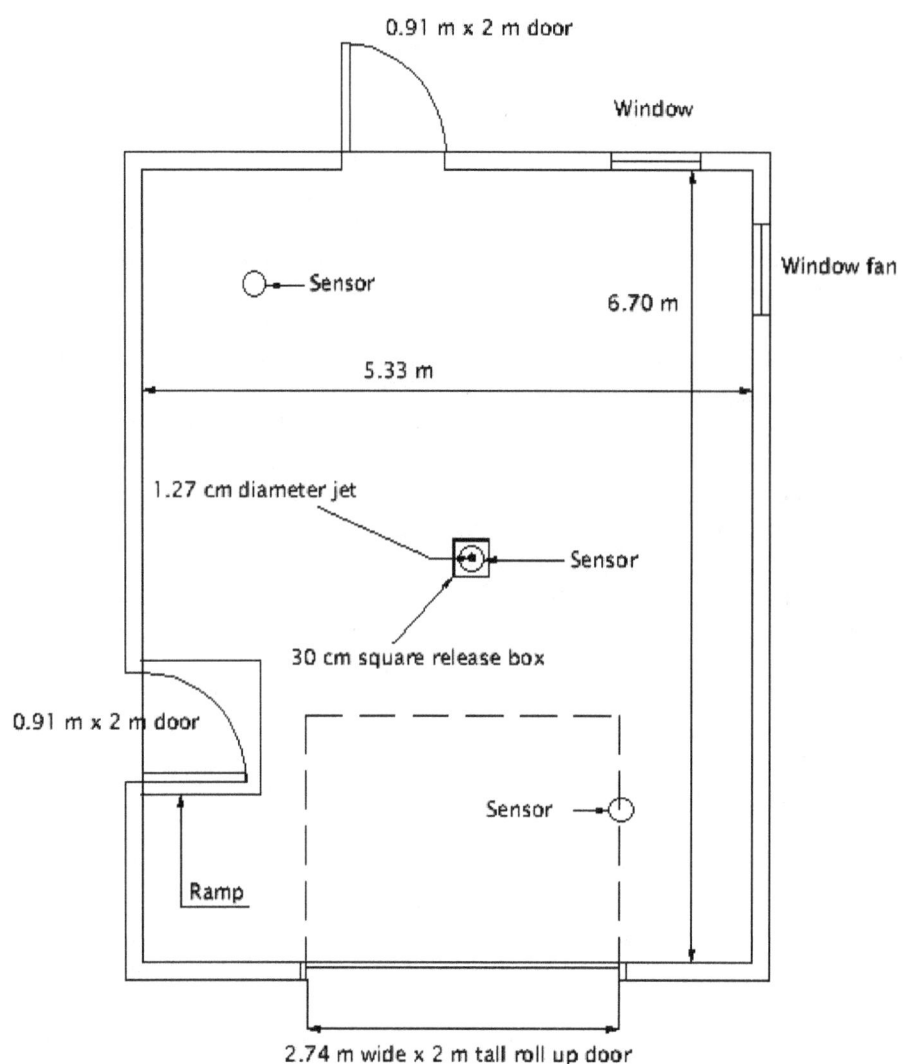

0.91 m x 2 m door

Window

Window fan

6.70 m

Sensor

5.33 m

1.27 cm diameter jet

Sensor

30 cm square release box

0.91 m x 2 m door

Sensor

Ramp

2.74 m wide x 2 m tall roll up door

Figure 3. This schematic shows a floor plan for the garage attached to the Indoor Environment and Ventilation Test House, Building 423.

Measurements of ACH for the garage were made using both tracer gas methods and doorway fan pressurization tests during related experiments carried out in conjunction with the measurements discussed in this report. Prasad et al. have described the results of a series of doorway fan tests carried out in the garage. [8] For measurements with the fan located in the rear door frame, values of $(ACH_{gar})_{4\ Pa}$ = 2.5 h^{-1} and 3.6 h^{-1} were measured for pressurization and depressurization tests, respectively. The corresponding pressure exponents, n (see Eq. (2)), were 0.50 and 0.61. The n values suggest air exchange was occurring through openings of sufficient size that turbulent flow could develop. [20] By using low levels of released hydrogen as a tracer gas, Cleary and Johnsson made several measurements of ACH_{gar}. [9] Their values ranged from 0.31 h^{-1} to 0.43 h^{-1}. The difference between $(ACH_{gar})_{4\ Pa}$ and ACH_{gar} reinforces the conclusion that effective differential pressures between the interior and surroundings for the garage were normally considerably smaller than 4 Pa. These values of ACH suggest that the NIST test garage is better sealed by roughly a factor of two than a "typical" garage reported in the literature, but the measurements are well within the range of values observed in earlier studies.

Figure 4. An interior view of the garage looking toward the front in which the floor and garage door can be seen.

2.3 Helium Supply and Release Systems

The helium control flow system was assembled in the family room (see Figure 2) of the house. Helium was supplied from a Scott Specialty Gases[1] Model 8403 Complete System Changeover Regulator System which combines a gas regulator with a two-way valve that allowed seamless switching between two different feed lines. Each of the two inlets, equipped with valves and bleed ports, was connected by tubing to short pressure manifolds equipped with two valves and gas lines running to two gas cylinders. A gas storage rack was used to safely hold the four gas storage bottles feeding the regulator system.

A typical experiment required twelve gas cylinders to provide the 10 kg of helium released over a 4 h period. The procedure employed was to supply helium from one of the manifolds fed by two high-pressure Type 1A cylinders until the pressure to the regulator dropped to roughly 350 kPa. At this point the flow was switched to the second set of gas cylinders, and the two empty cylinders were replaced by charged cylinders. This process was repeated until the end of the experiment.

The output of the regulator was connected through tubing to a valve leading to an American Meter Company Model AT-1400 diaphragm gas meter. The valve was used to control the flow rate. A pressure gauge was included to determine the gauge pressure at the meter. The gas flow rate was determined by timing the period required for 0.142 m^3 of helium to pass through the meter with a digital stopwatch. The gas pressure was measured at the same time. Volume flow rate measurements were recorded periodically (typically every 10 min) while the helium was flowing. Since the gas was released at atmospheric pressure, the measured volume flow rate was corrected to atmospheric pressure using the measured gauge pressure at the meter. The helium flowed into the garage through polyethylene tubing that passed through a duct placed in the wall between the family room and the garage.

---

[1] Certain commercial equipment, instruments, or materials are identified in this paper to foster understanding. Such identification does not imply recommendation or endorsement by the National Institute of Standards and Technology, nor does it imply that the materials or equipment identified are necessarily the best available for the purpose.

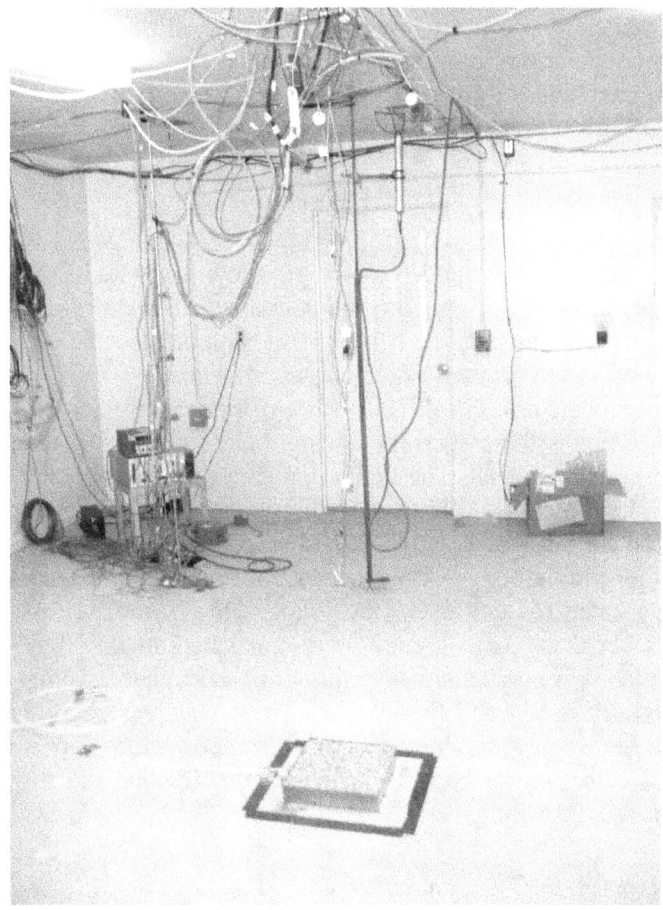

Figure 5. An interior view of the garage looking toward the rear is shown. The exterior doorway on the rear wall can be seen. The helium release flow conditioner used for the 2010 experiments is visible on the floor in the foreground.

The diaphragm test meter is a volume displacement measurement. The uncertainty in displaced volume was estimated to be 0.2 %. Uncertainties for the recorded times were estimated as 0.3 s and 700 Pa for the gauge pressure. A propagation of error calculation for the volume flow rate uncertainty gave an expanded uncertainty, i.e., coverage factor of 2, of $\pm\, 1.2 \times 10^{-4}$ m^3/s.

The helium was released into the garage in two different ways. In the 2008 experiments, the helium was released through polyethylene tubing (I. D. = 6.35 mm, O. D. = 9.53 mm). The tube was attached along a vertical mount at the center of the garage (x,y) = (2.69 m, 3.40 m). The helium flow exited the tube vertically upwards at a height of 41.5 cm $\pm$ 0.1 cm above the floor. For a nominal volume flow rate of 4.14 L/s, the corresponding exit flow velocity was 131 m/s. Such flows are typically characterized by the Reynolds number,

$$Re = \frac{\rho_{He} U_o D_o}{\mu_{He}}, \tag{3}$$

where ρ_{He} is the density of the gas at the exit, U_o is the average velocity, D_o is the diameter of the flow exit, and μ_{He} is the dynamic viscosity for helium, and the Froude number,

$$Fr = \frac{\rho_{He}^{\frac{1}{2}} U_o}{(g D_o (\rho_{air} - \rho_{He}))^{\frac{1}{2}}}, \tag{4}$$

where g is the gravitational constant and ρ_{air} is the air density. The values for these flow conditions assuming a temperature of 20 °C were Re = 7070 and Fr = 210. The Re and Fr values indicate the initial flow was turbulent and dominated by inertial forces.

An important characteristic of flows of buoyant gases in enclosures is the downstream distance, z, where the transition from momentum- to buoyancy-dominated flow takes place. According to Chen and Rodi, the nondimensional parameter K, given by

$$K = \frac{1}{Fr\left(\dfrac{\rho_{He}}{\rho_{air}}\right)^{1/4}} \frac{z}{D_o},$$

(5)

is < 0.5 for momentum-dominated flows and > 5 for buoyancy-dominated flows. [32] For the tube flow used in 2008, the flow began to transition from momentum- to buoyancy-dominated at $z \approx 1.9$ m and would have become buoyancy dominated at $z \approx 10.9$ m. This indicates that by the time the flow reached the ceiling (height of 2.43 m) it was primarily momentum dominated.

For the series of experiments done in 2010, a system was constructed to release helium uniformly over a much larger area. A box with an open top having dimensions of 30.5 cm × 30.5 cm × 6.4 cm was constructed from sealed thin sheet metal. The helium entered through one side of the box near the base and passed into a 3.8 cm high plenum. The top of the plenum was formed by a perforated metal sheet (diffuser) topped with a 10 cm layer of 12 mm crushed stone. The diffuser and stones were designed to smooth out the velocity profile before the flow exited upwards into the garage though the 30.5 cm × 30.5 cm square opening located 6.4 cm above the floor. This helium release system is visible in the foreground of Figure 5. For these experiments the nominal helium flow velocity into the garage was 0.045 m/s. The Reynolds and Froude numbers for this flow, calculated assuming a circular opening with the same release area, were $Re = 131$ and $Fr = 0.075$. These values indicate that the flow at the release location was laminar and buoyancy dominated. The distance required for the flow to become fully buoyancy dominated was calculated to be 0.21 m, which implies that the flow was a fully buoyancy-driven plume when it reached the ceiling.

2.4 Instrumentation

2.4.1 Helium sensors

Local helium volume percent was measured at multiple locations using Xensor Integration TGC-3880 thermal conductivity sensors. Approaches for calibrating these sensors and calculating volume percents were developed during the study of helium mixing and dispersion when released into a ¼-scale two-car garage. [1] Each sensor was covered with a standard 5 mm high × 8 mm diameter circular TO-5 cap with a 0.28 cm diameter hole drilled in the center that reduced its sensitivity to flow, while providing time responses on the order of a second. The sensor calibration and analysis procedures were identical to those employed in the earlier study. It was shown previously that for the range of helium volume percent in the current investigation, variations in measured helium concentrations between different sensors were less than 0.5 %, and estimated absolute uncertainties were less than 1 %.

Similar to the earlier reduced-scale study [1,2], eight sensors were used during the 2008 testing. The helium sensors were placed on mounts attached to a vertical steel pole located at $(x,y) = (1.37$ m, 5.09 m), which was close to the center of the left rear quadrant of the garage. The sensors and support pole are visible on the left side of Figure 5. The eight sensors were placed at heights of $z = 0.152$ m, 0.457 m, 0.762 m, 1.067 m, 1.372 m, 1.676 m, 1.981 m, and 2.286 m with an uncertainty of ± 0.002 m (coverage factor = 1). Table 1 lists the coordinates of the sensors for these tests.

For the 2010 tests, an additional six helium sensors were added for a total of fourteen. As discussed further below, experiments were performed with and without an automobile present in the garage. A vertical array of eight sensors was located at the same horizontal position as for the 2008 measurements. The sensor heights were $z = 0.305$ m, 0.610 m, 0.914 m, 1.219 m, 1.524 m, 1.829 m, 2.134 m, and

14

Table 1. Helium Sensor Locations for 2008 Tests

Sensor	x (m)	y (m)	z (m)
#1	1.37	5.09	0.152
#2	1.37	5.09	0.457
#3	1.37	5.09	0.762
#4	1.37	5.09	1.067
#5	1.37	5.09	1.372
#6	1.37	5.09	1.676
#7	1.37	5.09	1.981
#8	1.37	5.09	2.286

Table 2. Helium Sensor Locations for 2010 Tests without Vehicle

Sensor	x (m)	y (m)	z (m)
#1	1.37	5.09	0.305
#2	1.37	5.09	0.610
#3	1.37	5.09	0.914
#4	1.37	5.09	1.219
#5	1.37	5.09	152.4
#6	1.37	5.09	1.829
#7	1.37	5.09	2.134
#8	1.37	5.09	2.388
#9	1.37	5.09	0.051
#10	2.69	3.34	0.673
#11	2.69	3.34	1.219
#12	2.69	3.34	1.626
#13	2.69	3.34	2.134
#14	4.34	1.82	2.134

2.388 m. When an automobile was not present, a ninth sensor was added to the vertical array at z = 0.051 m. Four of the remaining sensors were centered over the helium release location, (x,y) = (2.69 m, 3.40 m), with heights of z = 0.673 m, 1.219 m, 1.626 m, and 2.134 m. The remaining sensor was placed 2.134 m above the floor in the front right quadrant of the garage at (x,y) = (4.34 m, 1.82 m). Table 2 lists the coordinates of the fourteen sensors for the 2010 experiments without a vehicle present.

As discussed in Section 2.5 below, either a Dodge Stratus or Volkswagen Passat was parked in the garage for some tests. When an automobile was present, Sensors #9 to #14 were moved to various locations underneath and inside the vehicle. Sensor #9 was placed in the driver's side front wheel well directly above the tire. Sensor #10 was attached to the dome light or rear view mirror in the center of the passenger compartment 6.4 cm from the ceiling. Sensor #11 was centered inside the trunk at mid height. Sensor #12 was placed inside the engine compartment in front of the engine near the base. Sensor #13 was placed at the top of the engine compartment. Sensor #14 was placed in the undercarriage of the vehicle. For the Stratus, a location centered crosswise in the area underneath the vehicle 61.0 cm in front (relative to the automobile) of the helium release system and 29.2 cm above the floor was used. Much of the underside of the Passat was protected by plastic splash shields which extended down to the level of the outer edge of the vehicle, and it was not possible to place the helium sensor in the same location as for the Stratus. As an alternative, Sensor #14 was placed on top of the left front wheel control arm. Table 3 lists the locations of Sensors #9 to #14 when the vehicles were present in the garage.

The voltages generated by the helium sensors were digitized using a SCXI 32-channel National Instruments 1102 signal conditioning board coupled with a SCXI-1300 interface board. The conditioned signal was fed to a PXI-6221 digitizer board. The data acquisition system was located in the garage near the sensors and connected by Ethernet to a personal computer located in the family room of the house.

Table 3. Locations of Sensors #9 to #14 inside Vehicle (when present) for 2010 Tests

Sensor	Location in Vehicle
#9	Front driver's side wheel well, 0.76 m above floor (Stratus), 0.74 m above floor (Passat)
#10	Attached to passenger compartment dome light at center, 6.4 cm from ceiling (Stratus), attached to rear view mirror at front 6.4 cm from ceiling (Passat)
#11	Centered in trunk, 24.1 cm above floor (Stratus), 31.8 above floor (Passat)
#12	Inside engine compartment, centered in front of engine 0.406 m above floor (Stratus), 0.46 m above floor (Passat)
#13	Inside upper engine compartment, centered left to right, 8.9 cm from hood and 0.305 m from rear gasket (Stratus), above engine on driver's side, 2.5 cm from hood, 25 cm from firewall and centerline (Passat)
#14	Centered crosswise in undercarriage, 0.61 m from front edge of helium release, 29.2 cm above floor (Stratus), attached above lower control arm on front driver's side , 18.4 cm above floor (Passat)

National Instruments LabVIEW software was used to control the data acquisition. Voltages were digitized at a 2 kHz rate and then averaged for either 1 s or 10 s before being saved in a comma-delimited file along with the relative time of the sample. It was possible to change the data acquisition rate during an experiment.

2.4.2 Pressure measurement

A MKS Baratron 1333 Pa range transducer was used to record differential pressures between the interior of the garage and the outside for some experiments. The transducer was calibrated in the same way described in [1]. The expanded uncertainty based on a linear least squares fit to the calibration data was ± 0.3 Pa. The pressure transducer was placed inside the garage. The voltage output was connected to the same digitization system used for the helium sensors. One of the pressure ports for the transducer was open, while the second was attached to a section of flexible tubing. The end of the tubing was passed underneath the rear door and exposed to the outside conditions near the base of the door. The pressure measurements tended to be very noisy due to wind turbulence. The measurements provided a qualitative indication of pressure fluctuations associated with the wind on the south side of the garage.

2.4.3 Helium measurements inside the test house

Helium volume percent levels inside the family room of the house were monitored at a single point using a Siemens Calomat 6 helium gas analyzer. The gas was sampled at a point in the center of the room at a height of 1.52 m by using a small pump to withdraw gas from the room into a tube and pass it continuously through the analyzer. A valve was used to set the flow rate through the instrument to the manufacturer's recommended value. The analyzer was calibrated by sequentially passing pure air and helium through the analyzer and adjusting the set points accordingly. The analyzer provided digital read out of helium volume fraction as well as an analog output. During the 2008 experiments, helium volume fraction was periodically read manually (roughly every 10 min) and recorded. Similar measurements were made for the second series of experiments, but, in addition, the analog output was connected to the data acquisition system and recorded automatically along with the outputs of the Xensor Integration sensors and the pressure transducer. The manual and recorded values were in good agreement. The manufacturer specifies the instrument to have an accuracy of 1 % helium or better over its full range.

2.4.4 Additional Temperature Measurements

Three Type K bead-type thermocouples were placed outside of the rear door for the garage, inside the garage near the side door, and inside the family room of the house in order to record exterior, interior garage, and interior house temperatures. The temperatures were recorded manually using portable

Figure 6. Photograph showing the test vehicle parked inside the garage.

thermocouple readers equipped with digital readouts. Temperatures were read at the same times that volume flow rate measurements and family room helium volume percents were recorded.

2.5 Experiments with a Vehicle Parked in the Garage

Experiments were performed with either a 2005 Dodge Stratus or a 2003 Volkswagen Passat parked in the garage in order to study the effect of a vehicle on helium mixing within the garage and to characterize the buildup of helium inside conventional vehicles. The nominal dimensions of these automobile were 1.79 m (w) × 4.85 m (l) × 1.37 m (h) for the Stratus and 1.74 m (w) × 4.70 m (l) × 1.46 m (h) for the Passat. The automobiles were centered over the 0.305 m × 0.305 m helium release system, i.e., centered in the garage. Figure 6 includes a photograph shot from outside through the garage door which shows the Stratus parked inside the garage. Experiments were run with all vehicle doors, windows, hood, and trunk closed.

2.6 Experiments with Forced Ventilation

The effectiveness of forced ventilation for limiting the buildup of helium in the garage was tested by performing experiments during the 2010 test series in which a vent was connected to an exhaust fan placed in the rear exit doorway of the garage. A 1.27 cm thick oriented strand board (OSB) wood panel was cut to fit inside the inner doorframe of the rear door and sealed into the frame using duct tape. A 10.8 cm diameter hole was cut in the panel centered 10.2 cm from the top of the doorway. A 20.3 cm × 20.3 cm square plate with a 10.8 cm diameter flange was held in place by screws and sealed over the hole on the outside using duct tape The flange was connected to an Infiltec Model DL1-DM4-110 duct leakage tester by a section of 10.8 cm diameter flexible plastic duct. This leakage tester is designed for use with residential heating and air conditioning systems. It provides a measured air volume flow rate while recording the pressure difference between the interior of the space and the volume on the downstream side of the fan. The device can provide calibrated volume flow rates ranging from

Figure 7. Photograph showing the OSB board with attached flange in place in the rear doorway of the garage. The flange was connected to the Infiltec Model DL1-DM4-110 duct leakage tester (visible at the base of the doorway) by flexible duct.

0.014 m^3/s to 0.146 m^3/s. The manufacturer reports the uncertainty in volume flow rate as 5 %. The fan was operated by drawing air from the garage, i.e., in depressurization mode. Figure 7 shows a photograph of the arrangement from the rear of the garage.

Six measurements employing forced ventilation were performed. All were run with automobiles parked centered over the helium release flow conditioner. Four of the experiments were done with the 2005 Dodge Stratus and two with the 2003 Volkswagen Passat.

The experiments were designed to mimic an automatically started fan responding to the detection of hydrogen by a sensor located near the ceiling. The fan on the Infiltec was started manually when the voltage for Sensor #8 (2.388 m above the floor) on the vertical array dropped to a voltage corresponding to a helium volume fraction of 1 %, which is ¼ of the lower flammability limit for hydrogen in air.

3   Experimental Results

3.1   Experiments with Natural Ventilation of the Garage

3.1.1   Helium release from a vertical tube without vehicle

Six experiments were performed with natural ventilation of the garage during the two series of experiments. Table 4 list the major experimental parameters including the date of the test, the type of helium flow, the average helium volume flow rate, the total helium volume flow, and whether or not a vehicle was present.

Figure 8 shows a plot of measured helium volume flow rate as a function of time for the 9/11/08 experiment. The helium was released upwards from the 0.635 cm diameter vertical tube. The helium flow was initiated at 120 s and halted at 13 710 s as the helium in the twelve cylinders was fully depleted. The total flow time was about 800 s shorter than the 4 h target time.

Table 4. Summary of Natural-Ventilation Garage Experimental Flow Parameters

Date	Helium Flow	Average Flow Rate (m^3/s)	Flow Period (s)	Total Flow Volume (m^3)	Vehicle?
9/11/08	tube	4.44×10^{-3}	13 590	60.3	No
9/12/08	tube	4.43×10^{-3}	12 840	56.9	No
7/29/10	box	4.30×10^{-3}	14 352	62.1	No
8/2/10	box	4.36×10^{-3}	14 400	62.7	No
8/6/10	box	4.47×10^{-3}	14 400	64.3	Yes
8/23/10	box	4.32×10^{-3}	13 636	59.1	Yes

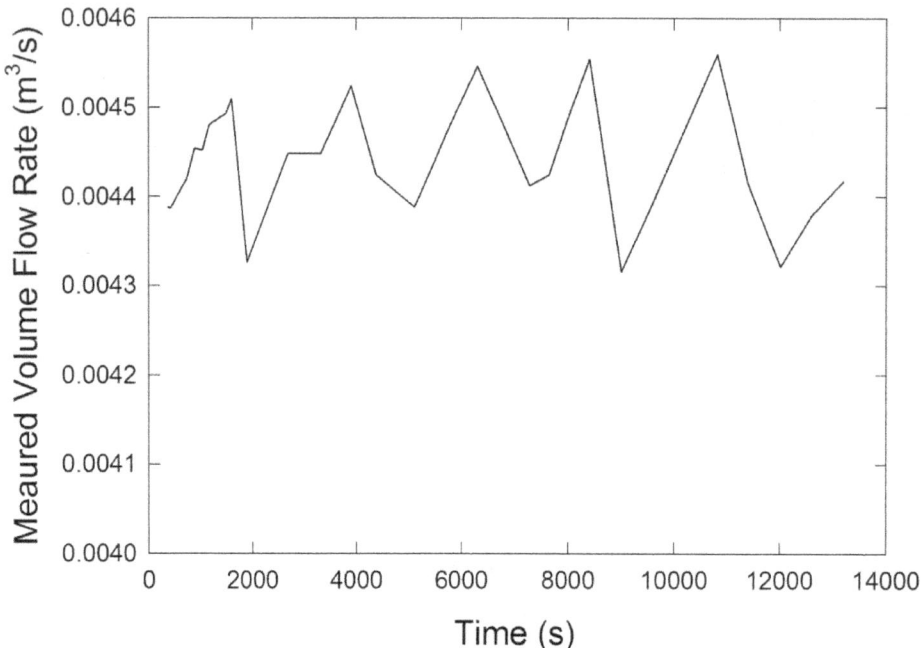

Figure 8. Measured helium volume flow rate is plotted as a function of time for the 9/11/08 garage experiment. Measurements were recorded roughly every 600 s.

A small repetitive fluctuation of the volume flow rate is evident in Figure 8. These fluctuations were roughly correlated with the times when the switchovers between pairs of helium cylinders were made, suggesting a weak sensitivity of the regulated flow to cylinder pressure or other property of the cylinders.

The average helium volume flow rate for the data shown in Figure 8 was 4.44×10^{-3} m^3/s with a variance of 6×10^{-5} m^3/s. This flow rate is 7 % higher than the target value of 4.15×10^{-3} m^3/s. The total volume of helium released was estimated by performing a piecewise time integration over the measurements. The result was 60.3 m^3, which was 1 % higher than the target value of 59.7 m^3, even though the release time was less than 4 h.

Figure 9 shows measured helium volume percent for the eight sensor locations along the vertical array as a function of time for the 9/11/08 experiment. A period covering 12 h is shown. Several distinct behaviors are evident. The helium concentrations at all heights began to increase shortly after the flow was started at 120 s. This is evident in Figure 10, which shows the concentration profiles during the initial 1000 s of the experiment. The initial detection of helium at Sensor #8 occurred at 136 s or 16 s after the flow started. The 16 s represents the time required for helium to flow to the release point, rise to the ceiling, and spread out to reach the sensor location. The helium then moved downward from the

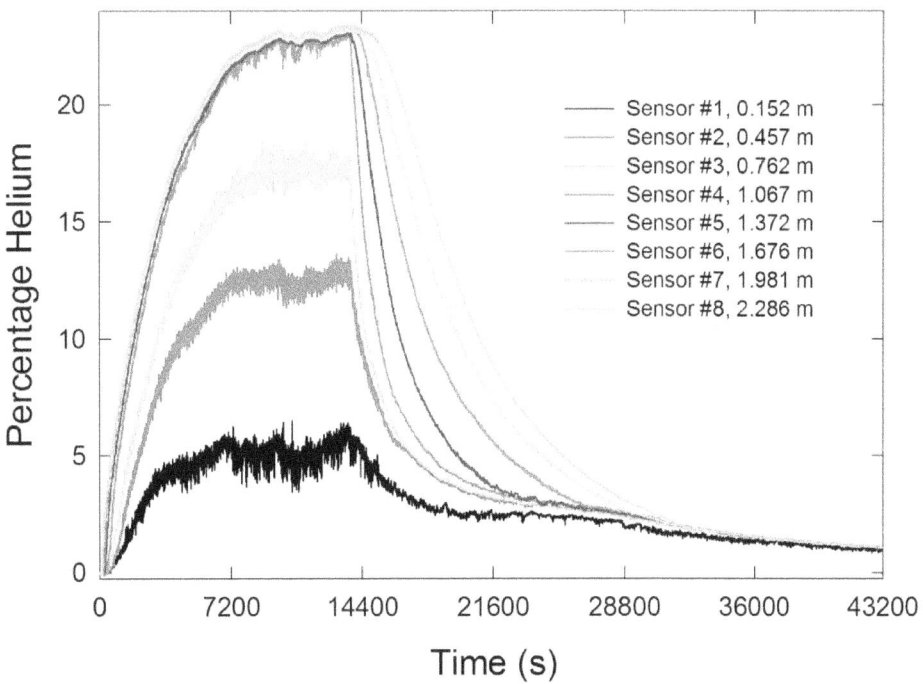

Figure 9. Helium volume percent is plotted as a function of time for the eight sensor heights indicated for the 9/11/08 experiment.

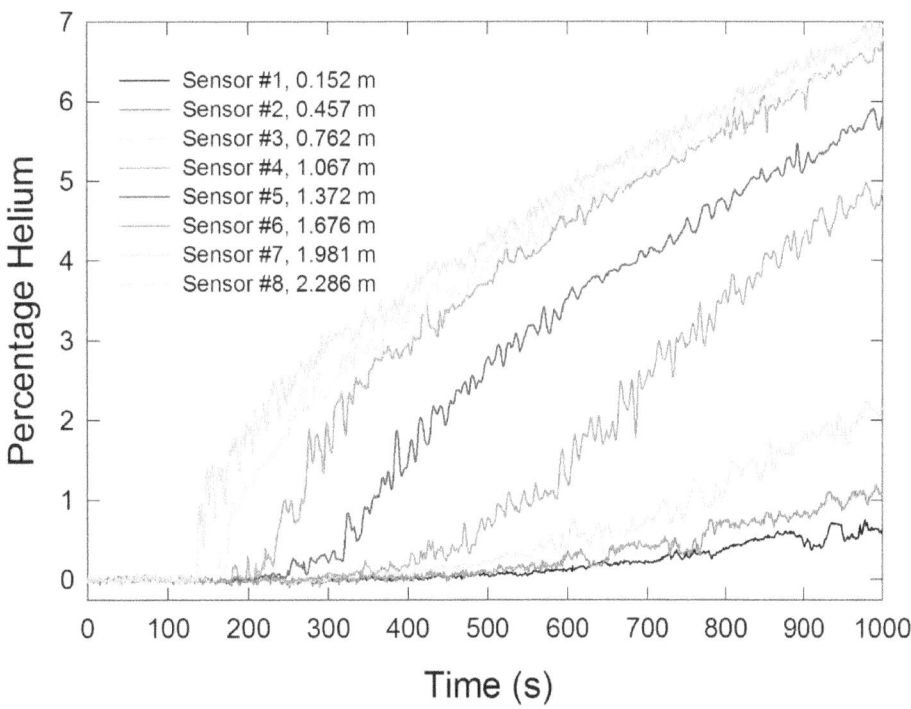

Figure 10. Helium volume percent is plotted as a function of time for the eight sensor heights indicated for the initial 1000 s of the 9/11/08 experiment.

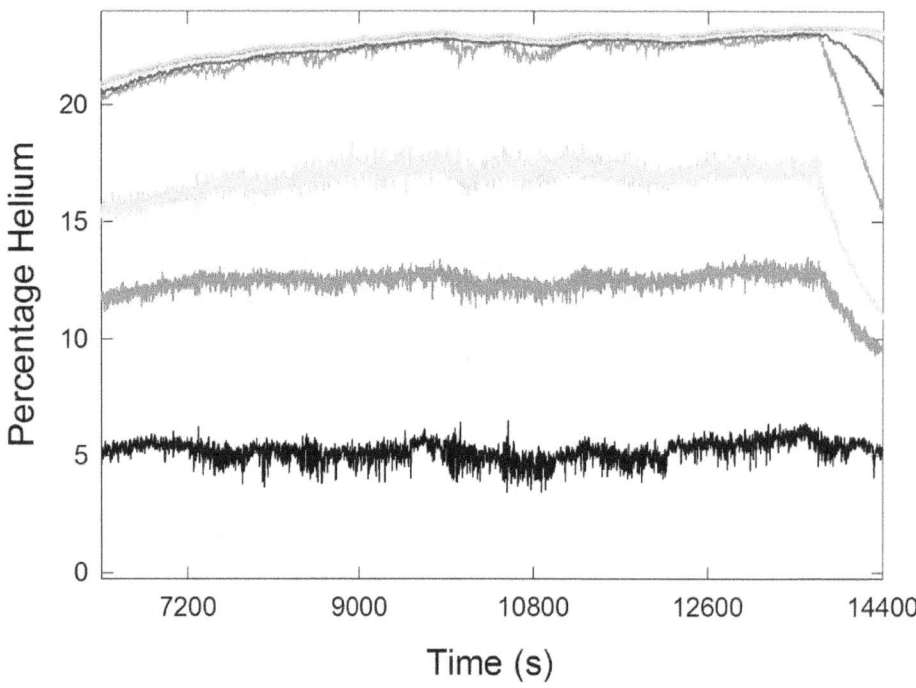

Figure 11. Helium volume percent is plotted as a function of time for the eight sensor heights indicated over the period from 6300 s to just after the end of the helium release for the 9/11/08 experiment. The color codes for the sensor locations are the same as in Figure 10.

ceiling, reaching the lower sensors sequentially from top to bottom. Only about 300 s from the start of the flow was required for helium to reach the lowest sensor.

In Figure 9, it is evident that a well mixed layer rapidly developed in the upper portion of the garage at heights of 1.1 m and higher and remained well mixed as the helium concentration increased. At the lower measurement points there was a rapid falloff in helium concentration with height that was evident over the entire release period. After roughly two hours the helium concentrations in the garage approached constant values for all of the measurement heights, indicating a quasi-steady state was developing. The well mixed upper layer and the concentration stratification at lower levels were present when the helium flow was halted.

Close inspection shows that the concentration profiles at the lower sensor positions had distinct rapid fluctuations (recall data were recorded at 1 Hz), while those in the well mixed upper layer were considerably smoother. This "unmixedness" resulted from significant concentration variations over relatively small distances, an indication of significant ongoing mixing, in the lower part of the garage volume. Note that the helium volume percent measurements at $z = 1.067$ m showed some fluctuations even though the concentration was very close to that of the well mixed upper layer. This provides evidence that the height of the interface between the well mixed upper layer and the stratified lower layer varied with time.

The helium volume percents recorded during the quasi-steady-state are shown on an expanded time scale in Figure 11. Additional correlated, lower frequency helium concentration variations are evident at all sensor heights. This suggests that an experimental parameter capable of affecting the quasi-steady state concentration levels was varying slowly over this period. One possibility is that variations in wind speed and direction altered the gas exchange rate between the garage and its surroundings. Figure 12 shows time plots of wind speed recorded by the sonic anemometer located 10 m above the ground and the differential pressure measured between locations outside the garage rear door and garage interior between 6 300 s and 14 400 s. Wind direction measurements showed that the wind was primarily from the south during this time. Even though these measurements were recorded at different locations using different

22

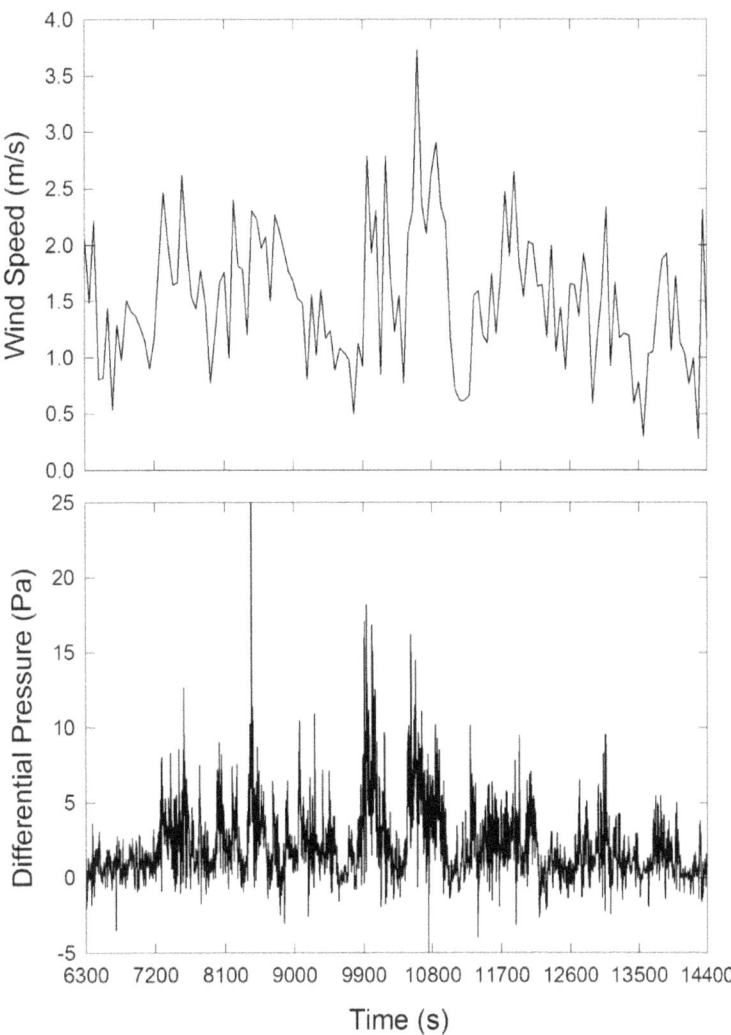

Figure 12. Time variations of wind velocity and the differential pressure between the garage interior and outside ambience are plotted over the time period from 6300 s to 14400 s for the 9/11/08 experiment.

data rates, visual comparison reveals that there was a strong correlation between wind velocity and the differential pressure, with periods of higher wind velocity corresponding to periods of higher differential pressure. Comparison of the data in Figure 12 with the helium concentration data in Figure 11 confirms that small dips in the quasi-steady-state concentrations were correlated with periods when wind velocity and differential pressure were higher. Apparently, the exterior weather condition had a small, but measurable, effect on the quasi-steady-state helium volume percents inside the garage.

The development of a quasi-steady state allowed a number of parameters characterizing the helium release phase to be defined. The first is the upper-layer helium volume percent at the end of the release period, which was determined to be 23.1 %. It was also possible to estimate the average concentration in the garage by assuming the absence of horizontal concentration gradients and extrapolating the measurements along the vertical array to the floor and ceiling. Justification for this assumption is provided below. Figure 13 shows an example of the approach for the 9/11/08 experiment. The experimental quasi-steady-state helium volume percents at the various sensor heights were determined by visual inspection of the plots shown in Figure 9 and are plotted in Figure 13. The measurements reveal that the vertical concentration profile can be fit well by two straight lines describing the constant

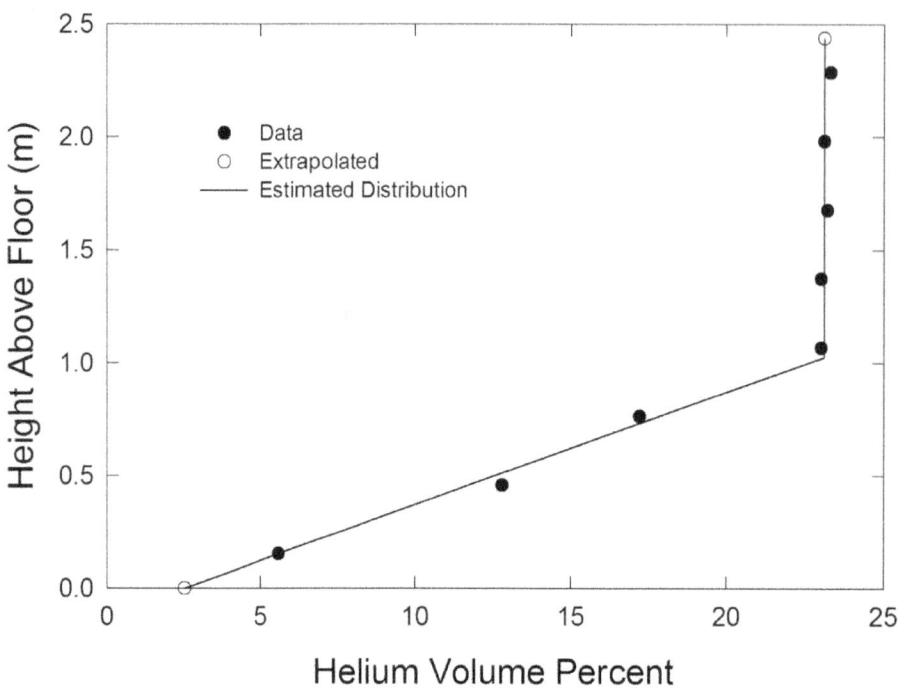

Figure 13. Experimental helium volume percents measured along the vertical array at the end of the 9/11/08 release are shown along with extrapolated values at the floor and ceiling. The solid lines are approximations of the data that are used to estimate the average helium concentration along the vertical direction.

Table 5. Helium Concentration Parameters for Garage at the End of Helium Release

Date	Upper Layer He Vol. % at End of Release	Average He Vol. % in Garage at End of Release	Volume of Released He in Garage at End of Release (m³)	Fraction of Released He in Garage at End of Release
9/11/08	23.1 % ± 0.3 %	18.8	16.3	0.27
9/12/08	23.1 % ± 0.3 %	19.0	16.5	0.29
7/29/10	22.3 % ± 0.3 %	19.5	16.9	0.31
8/2/10	22.7 % ± 0.3 %	20.2	17.6	0.28
8/6/10	22.3 % ± 0.3 %	20.2	17.6	0.27
8/23/10	22.1 % ± 0.3 %	20.1	17.5	0.30

concentration in the upper layer and the falloff with height in the lower layer. The lines allow extrapolation of values to the ceiling and floor as indicated by the open symbols in the plot. The upper-layer helium volume percent, the floor and ceiling extrapolated values, and the height where the lines intersected were varied qualitatively until the observed agreement between the idealized lines and the experimental data was achieved. The two lines were used to integrate the concentration along the height to obtain the average value for the helium volume percent. The result was 18.8 %.

By using the estimated average helium volume percent and the known volume of the garage (86.9 m³), the volume of helium present in the garage at the end of a release could be calculated. For the 9/11/08 experiment, the result was 16.3 m³. The fraction of released helium remaining in the garage could now be determined by dividing by the total released volume from Table 4 to give 0.27. This value along with the upper-layer helium volume percent and the average helium volume percent at the end of the release period is included in Table 5, which summarizes these values for the six experiments.

24

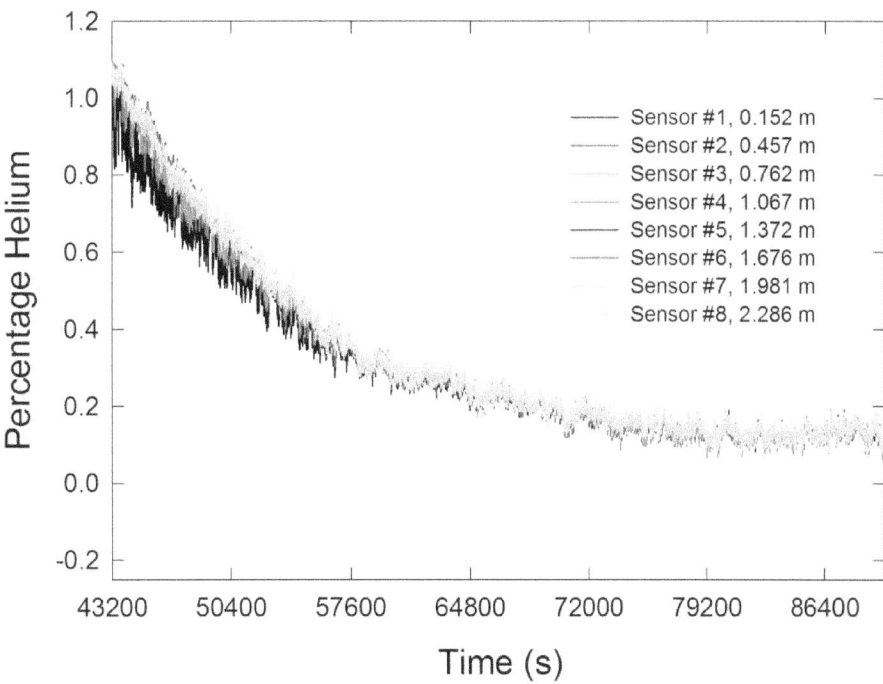

Figure 14. Helium volume percent is plotted as a function of time at the eight sensor heights indicated for the period from 43 200 s to 90 000 s in the 9/11/08 experiment.

In Figure 9 and Figure 11 it can be seen that the helium concentrations at the heights of Sensors #1 to #4 began to fall within seconds of the end of the helium release at 13 710 s. The most rapid initial drop was for Sensor #4, located at a height of 1.067 m, which was shown above to have been located close to the interface of the well mixed upper layer and the lower volume of the garage where a strong vertical concentration gradient developed. As time passed, the falloff in concentration moved upward through the upper layer, reaching the upper sensor location 2.286 m above the floor roughly 1500 s after the helium flow was stopped. Note that fluctuations in the quasi-steady state concentrations introduce some uncertainty, on the order of ± 200 s, in this estimate,

The helium concentration at each measurement location fell with a different rate in such a way that all of the profiles collapsed sequentially to a common curve from bottom to top. It is straightforward to estimate the times when concentration readings for a given sensor collapsed onto the common curve. The period required for the uniform concentration of helium to develop within the garage can be approximated as the time from the end of the release until the helium volume percent at Sensor #8 collapsed onto the common curve. For the data shown in Figure 9, roughly 18 000 s was required. At this time the helium volume percent was nearly uniform throughout the garage at ≈ 2.5 %. Interestingly, during the period when the collapse to a uniform helium concentration was occurring, the value for the lowest sensor at a height of 0.152 m remained nearly constant as the helium volume percents for the higher sensors approached this value sequentially from Sensor #2 to Sensor #8.

When the helium concentrations in the garage at the eight vertical sensors collapsed to a common decay curve, the falloff became relatively slow. Figure 14 shows the helium concentration decays for the period from 43 200 s to 90 000 s. The helium concentrations dropped from around 1 % to 0.2 % over this period. By assuming the helium was acting as a tracer gas, it was possible to estimate a value for ACH_{gar} by curve fitting an exponential curve to the falloff. The data for Sensor #4 was fit using a nonlinear least squares curve fit transform in SigmaPlot. Figure 15 shows the experimental data and the result for the fit. The calculated time constant was 7.47×10^{-5} s^{-1}, which corresponds to $ACH_{gar} = 0.27$ h^{-1}. The R^2 value for the fit was 0.973, indicating that an exponential curve was a good representation for the concentration decay.

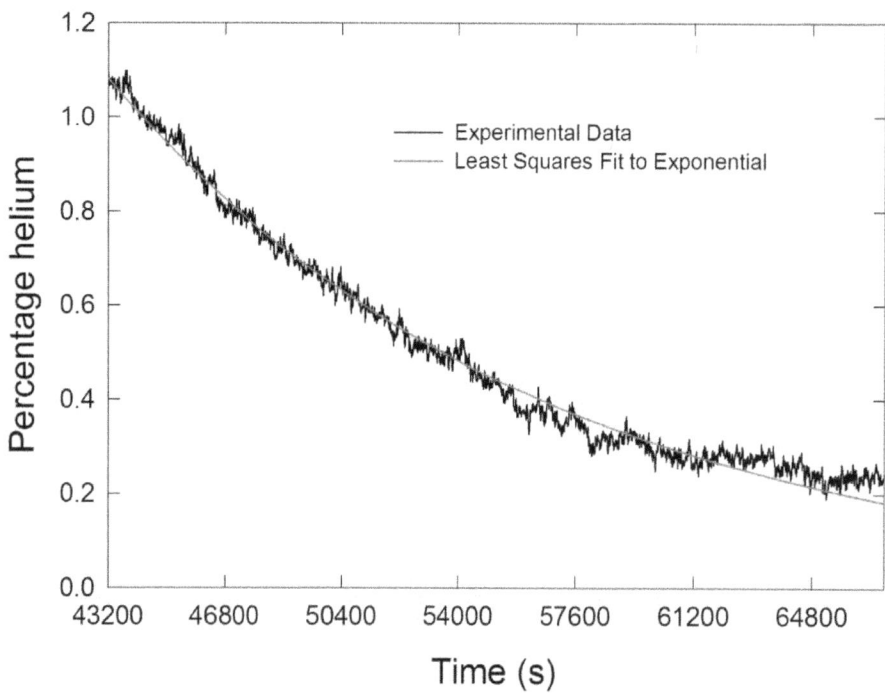

Figure 15. Helium volume percent recorded by Sensor #4 is plotted as a function of time for the 9/11/08 experiment. The red line is the result of fitting the data to an exponential decay using a nonlinear least squares curve fit.

Table 6. Garage Parameters for Post-Release Period

Date	Highest Sensor (m)	Time for Sensor #8 Initial Roll Off (s)	Time for Sensor #8 Collapse (s)	He Vol % at Sensor #8 Collapse	Concentration Decay Constant (s^{-1})	ACH_{gar} (h^{-1})
9/11/08	2.286	1 500 ± 200	18 000	2.5	7.47 × 10^{-5}	0.27
9/12/08	2.286	1 200 ± 200	15 200	3.6	3.14 × 10^{-5}	0.11
7/29/10	2.388	800 ± 200	12 400	3.0	9.08 × 10^{-5}	0.33
8/2/10	2.388	1100 ± 200	14 200	2.2	6.04 × 10^{-5}	0.22
8/6/10	2.388	900 ± 200	14 100	2.4	5.53 × 10^{-5}	0.20
8/23/10	2.388	900 ± 200	14 100	2.2	4.46 × 10^{-5}	0.17

The behavior of the helium concentration in the garage following the end of the helium release can be characterized in terms of the parameters defined above. Values for the periods between the end of the release and the initial falloff of concentration and for the collapse to the uniform concentration for the highest sensor, Sensor #8, at 2.286 m, the helium volume percent at collapse, and the decay rate at long times with associated value of ACH_{gar} are summarized in Table 6 for the 9/11/08 data. Values for later experiments discussed below are also included.

During the experiment, temperatures were measured in the family room of the house and outside using the dedicated thermistors and added thermocouples. Figure 16 compares the two sets of measurements. The measurements in the family room agreed very well when the different rates used to record data are taken into consideration. The distinct fluctuations in temperature resulted from the cycling of the air conditioning system used to cool the house. The temperature measurements outside do not agree as well, with the thermocouple measurements generally being somewhat higher. There are two possible reasons for this. The temperature may have been slightly higher just above the concrete outside the rear door of the garage where the thermocouple was places as opposed to the open air higher up where

26

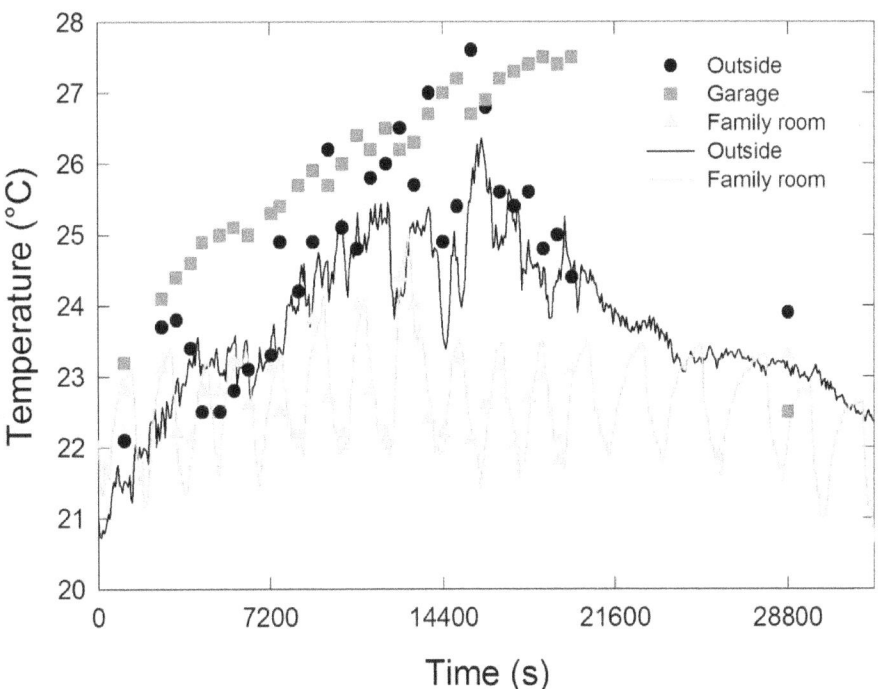

Figure 16. Temperatures measured using thermocouples (symbols) and thermistors (lines) are plotted as a function of time for measurements recorded on 9/11/08 in the family room, garage, and outside.

the thermistor was located. It is also possible that the thermocouple was not adequately shielded from radiation from the sun and sky and was radiatively heated. Figure 16 also includes temperature measurements made inside the garage using a thermocouple. Since this thermocouple was placed at an interior location, it would not have been subject to radiative heating. The temperature inside the garage rose during the helium release in a way that roughly tracked the outside temperature, even though the rate was a bit higher. As a result, the temperature difference between the interiors of the house and garage increased from about 1 °C to 5 °C during the helium release.

Figure 17 shows a plot of measured helium volume percent in the family room of the house as a function of time during and immediately following the helium release inside the garage. The helium concentration increased continuously, reaching a value of about 2.5 % shortly after 12 000 s, when a sudden drop occurred. This concentration drop happened when a window in the family room was opened around 12 300 s. The window remained open until 13 620 s. When the window was shut, the helium concentration quickly started to rise again, reaching values nearly as high as before the window was opened. The helium concentration remained relatively constant after the helium flow into the garage was halted at 13 710 s, only starting to decrease at longer times.

During the experiment, the fan for the air conditioning was on continuously as confirmed by the time record of the fan status recorded by the monitoring system for the house. Nabinger and Persily found that with the fan operating, a tracer gas was fully mixed throughout the house within ten minutes. [30] While helium at the levels observed in the family room may not act purely as a tracer gas, it is likely that it was mixed throughout the house volume within a similar time frame. This suggests that the helium volume percent measured in the family room was representative of a well mixed helium/air mixture throughout the interior volume of the house.

By assuming that the helium inside the house was fully mixed, it was possible to estimate the total volume of helium present inside the house at the end of the helium release in the garage by multiplying the measured volume percent, 2.3 %, at this time by the interior volume of the house. The result for the

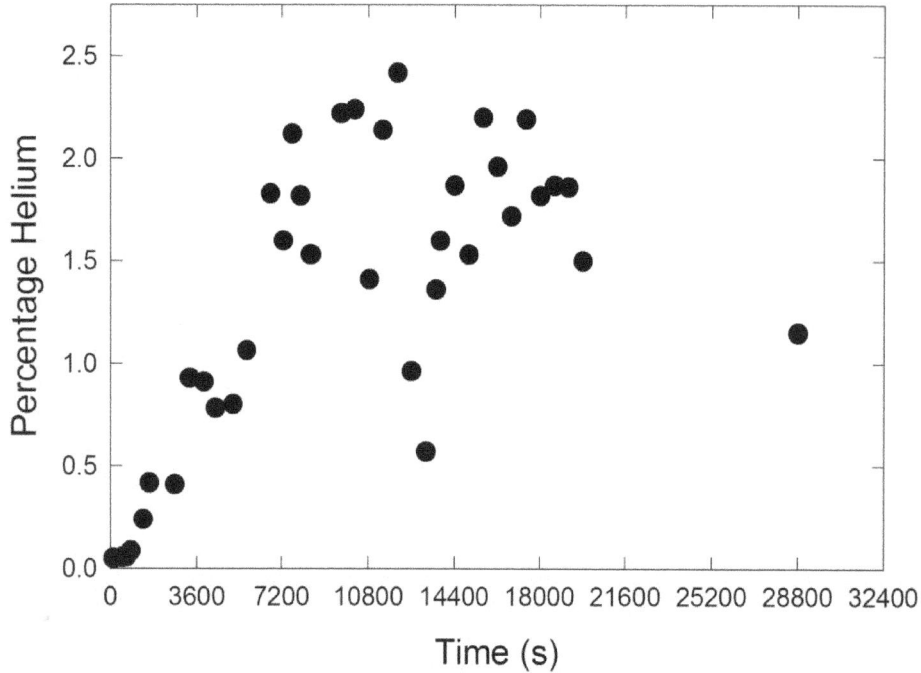

Figure 17. The helium volume percent measured in the family room of the house is shown as a function of time for the 9/11/08 experiment.

Table 7. Helium Parameters in the House at the End of Helium Release in the Garage

Date	He Vol. % in House	Volume of He in House (m³)	Fraction of Released He
9/11/08	2.3	7.8	0.13
9/12/08	4.8	16.3	0.29
7/29/10	3.7	12.6	0.20
8/2/10	4.1	13.6	0.22
8/6/10	3.2	10.9	0.17
8/23/10	2.5	8.5	0.14

9/11/08 data was 7.8 m³, which corresponds to 13 % of the total helium volume released into the garage. The helium volume percent in the house, corresponding total helium volume in the house, and the fraction of the total volume of helium released present in the house at the end of helium flow into the garage are summarized in Table 7.

The experiment with helium release from the straight tube was repeated on 9/12/08. The measured helium volume flow rate had a similar variation with time to that shown in Figure 8 for the 9/11/08 experiment. The average helium volume flow rate, flow period, and the total volume released for the 9/12/08 experiment are included in Table 4. The average helium volume flow rates were similar for the two experiments. However, since the release period for the second experiment was shorter, the total volume of helium released was reduced by 5.6 %.

Figure 18 shows a plot of measured helium volume percent as a function of time at the eight sensor locations for the 9/12/08 experiment. The general time dependence of the concentration profiles are similar to those for the 9/11/08 case shown in Figure 9. The uniform upper layer helium volume percent at the end of the release period was 23.1 % (see Table 5), which is the same as observed on 9/11/08. Close inspection reveals that the helium concentrations for each of the sensors in the lower part of the garage were slightly higher on 9/12/08. As a result, when the experimental data was fit and integrated

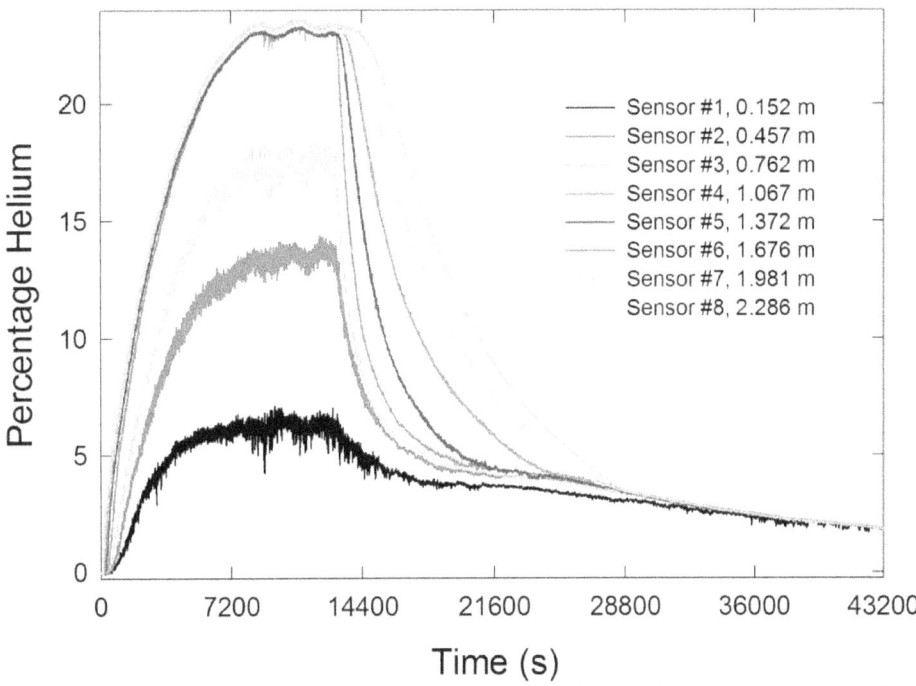

Figure 18. Helium volume percent is plotted as a function of time at the eight sensor heights indicated for the 9/12/08 experiment.

using the same approach as above to determine a quasi-steady-state average helium volume percent at the end of the release, the value was slightly higher for the second set of data. While concentration fluctuations are evident in the quasi-steady-state portion of the helium release for the second experiment in Figure 18, they were not as intense as for the original data.

The wind velocity and differential pressure between the garage and outside ambience are shown in Figure 19 during the quasi-steady-state release portion of the 9/12/08 data. During this time the wind was primarily from the south. As observed for the 9/11/08 experiment, there was a clear correlation between the strength and fluctuations of the wind velocity and the differential pressure fluctuations. Comparison with Figure 18 shows that dips in the measured helium concentrations and the strength of the fluctuations were anti-correlated with the wind velocity and differential pressure magnitudes as was the case for the earlier experiment. Figure 12 and Figure 19 show that both the wind speed and differential pressure fluctuations were higher on 9/11/08. Assuming gas exchange rates between the garage and its surroundings increase with higher wind speeds and differential pressures, this observation provides a plausible explanation for the slightly higher average helium volume percent at the end of the release observed on 9/12/08.

Following the helium release, the concentrations fell off in similar ways for both experiments, but, as evident from the results listed in Table 6, the time required for the concentration profiles to collapse was shorter, and the helium volume percent when all of the curves collapsed was higher for the second experiment. Comparison of the concentration profiles for the two experiments reveals that their decay behaviors were similar. The principal reason for the faster collapse of the curves on 9/12/08 was the higher helium volume percent when the single decay curve developed. A fit of the falloff data at long times when the helium volume percent was on the order of 1 % and lower yielded an estimate of ACH_{gar} = 0.11 h^{-1}, compared to ACH_{gar} = 0.27 h^{-1} for the earlier data. The experimental observations for the two experiments during the release and post-release periods are consistent, with slower air incursion into the lower level of the garage for the 9/12/08 experiment.

29

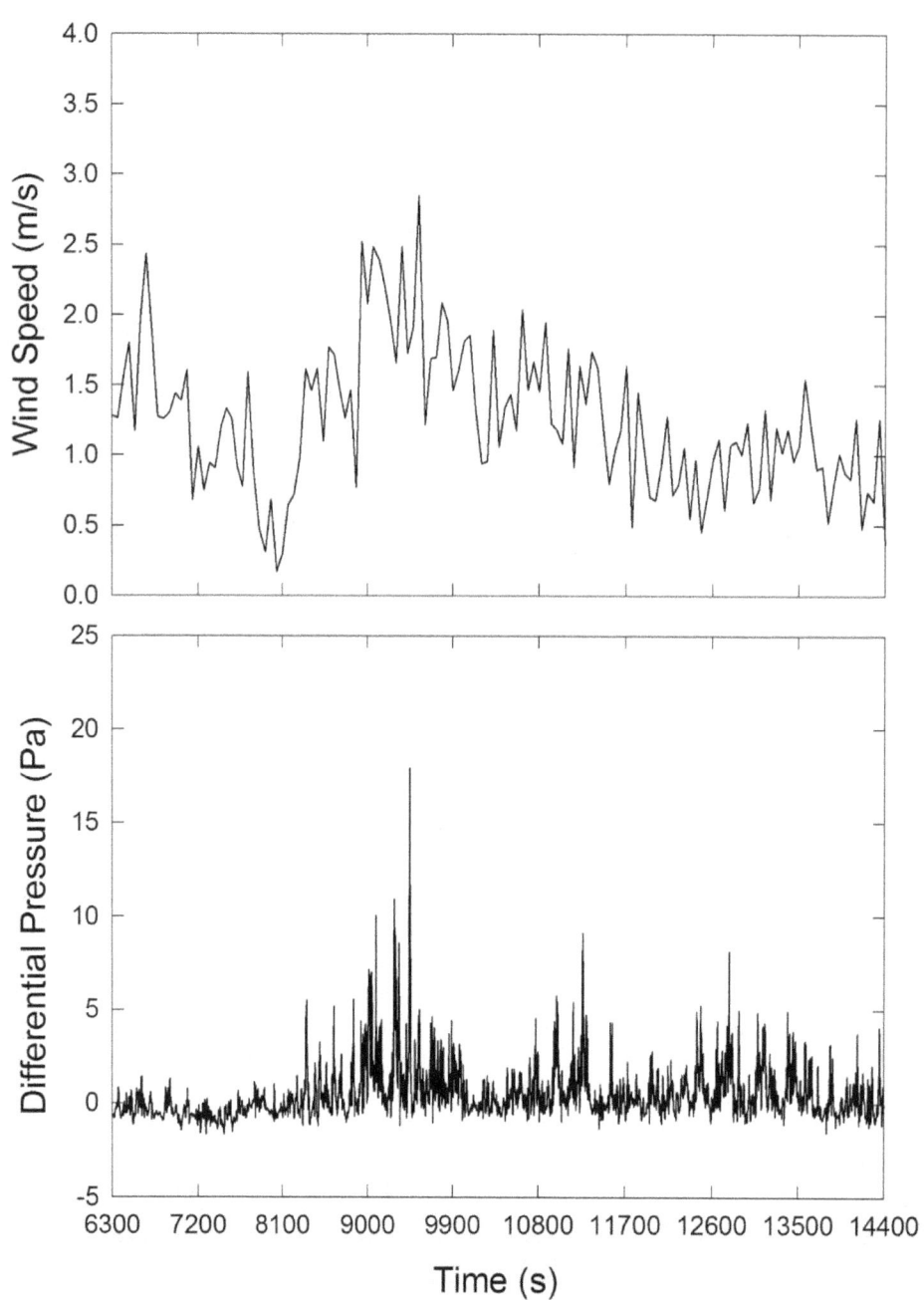

Figure 19. Time variations of wind velocity and the differential pressure between the garage interior and outside ambience are plotted over the time period from 6300 s to 14400 s for the 9/12/08 experiment.

Figure 20 shows comparisons of the wind speed and differential pressure between the garage interior and outside during the twelve hours following the start of the experiments on 9/11/08 and 9/12/08. The corresponding plots of helium volume percent over the same time periods are shown in Figure 9 and Figure 18. During both experiments the wind shifted from the south to primarily from the north shortly after the helium flow was halted. Over most of the release and immediate post-release periods the wind and differential pressure magnitudes and fluctuations were noticeably higher for the earlier experiment. If a higher wind is partially responsible for increased gas exchange with the garage, as expected, this provides an explanation for the higher rate of helium losses from the garage over these times.

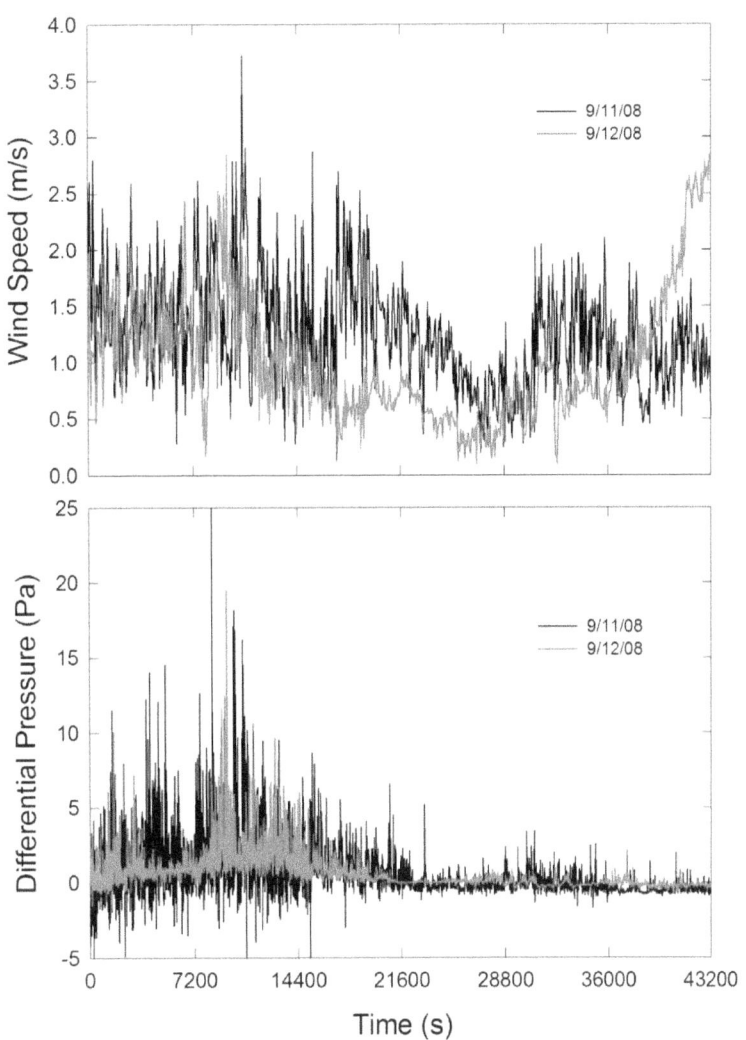

31

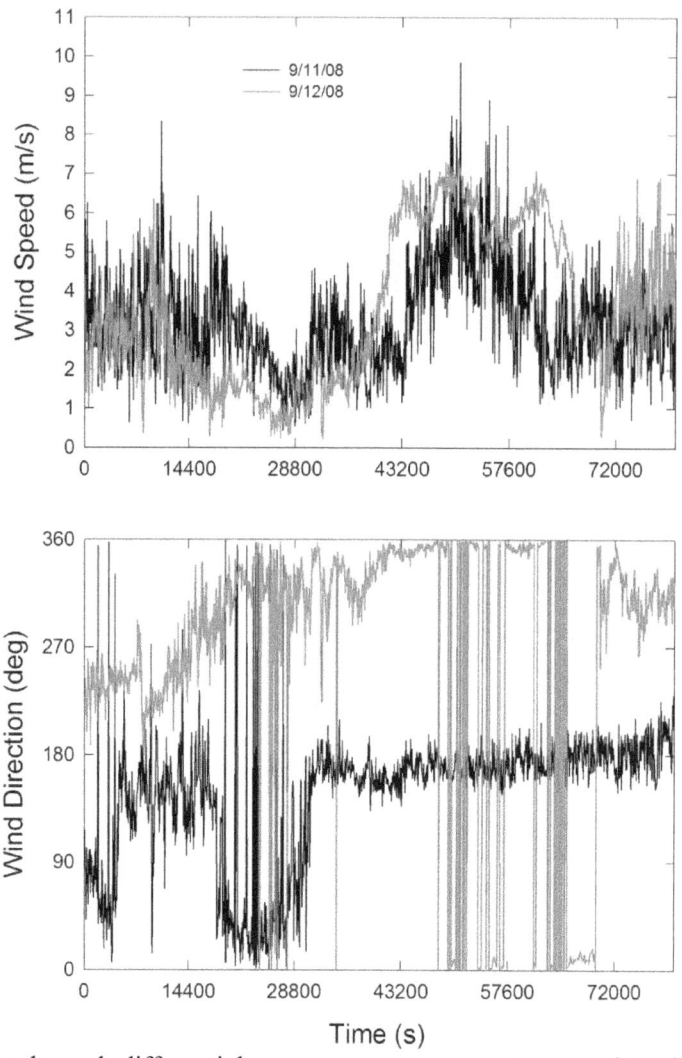

Figure 20. Wind speeds and differential pressure measurements are plotted over the twelve hours following the start of the experiments on 9/11/08 and 9/12/08.

A steep increase in wind velocity starting around 36 000 s for the 9/12/08 experiments was not reflected in the differential pressure measurement. The wind direction during this period was from the north, as opposed to measurements during the quasi-steady-state helium release period when the wind was from the south for both experiments. This suggests that the differential pressure measurements are wind-direction sensitive, with the largest response observed for winds blowing in the direction of the measurement location, i.e., from the south.

The importance of wind direction becomes clearer when considering differences in measured ACH_{gar} for the two experiments. Wind velocities and directions are compared in Figure 21 over longer times, which include the periods when ACH_{gar} values were determined. A value of 0° is defined to be north. At roughly 12 h (43 200 s) after the start of the experiment on 9/12/08, the wind began to blow strongly from the north, while on 9/11/08 the wind was from the south over the same period. Even though the wind magnitude at long times was higher, the value of ACH_{gar} for the 9/12/08 experiment was 2.5 times smaller than measured on 9/11/08. This is strong evidence that gas exchange between the garage and its surroundings was strongly dependent on wind direction as well as magnitude.

The weather during the 9/12/08 helium release was cool and cloudy with rain. These conditions were reflected in the temperature measurements shown in Figure 22 for locations in the family room, garage,

32

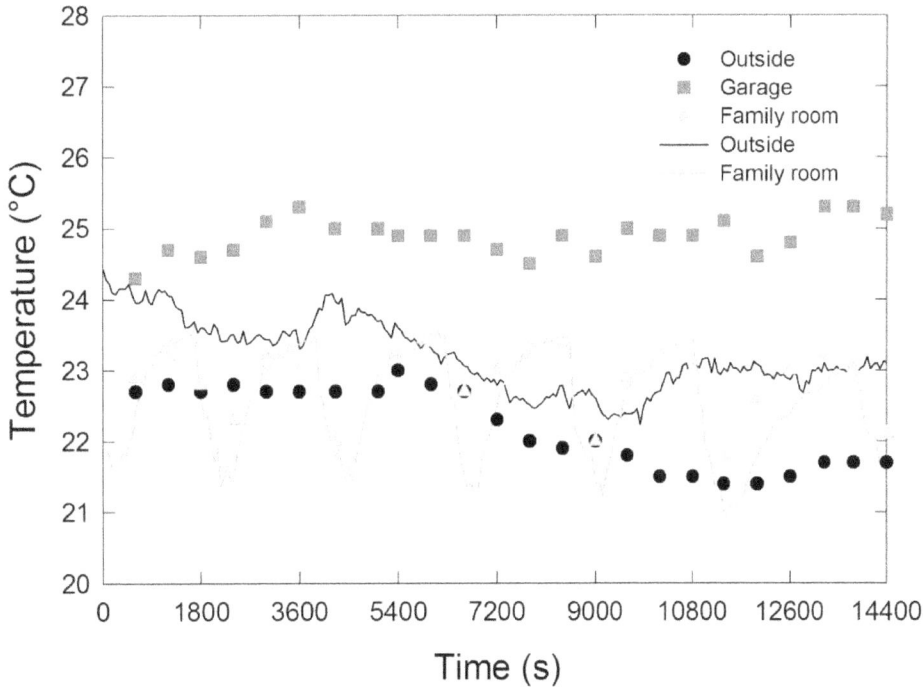

Figure 21. Wind speed and direction are plotted as a function of time for the 9/11/08 and 9/12/08 experiments.

and outside. There were relatively small differences between the locations. Unlike the similar measurements shown in Figure 16 for measurements on 9/11/08, outside temperatures recorded by the thermocouples were slightly lower than measured with the thermistor. This may suggest that interactions with a nearby surface played a role in the thermocouple measurements. The temperature fluctuations associated with the cycling of the air conditioning were evident in the family room. The fan was operated in continuous mode.

The helium volume percent recorded in the family room during the 9/12/08 helium release is shown in Figure 23. The concentration rose continuously over the period, approaching 4 % helium by the end of the release. As discussed above, since the fan was running, it is likely that the entire volume of the house reached a level around 4 %. This value along with derived values for the helium volume in the house at the end of the release and the fraction of the total released helium it represented are included in Table 7. All of the values were much larger than those measured on 9/11/08. Comparison of Figure 23 with the corresponding plot for the 9/11/08 data in Figure 17 shows that the helium concentrations roughly tracked each other until the window in the family room was opened during the earlier experiment.

Figure 22. Temperatures measured using thermocouples (symbols) and thermistors (lines) are plotted as a function of time for measurements recorded on 9/12/08 in the family room, garage, and outside.

Figure 23. The helium volume percent measured in the family room of the house is shown as a function of time for the 9/12/08 experiment.

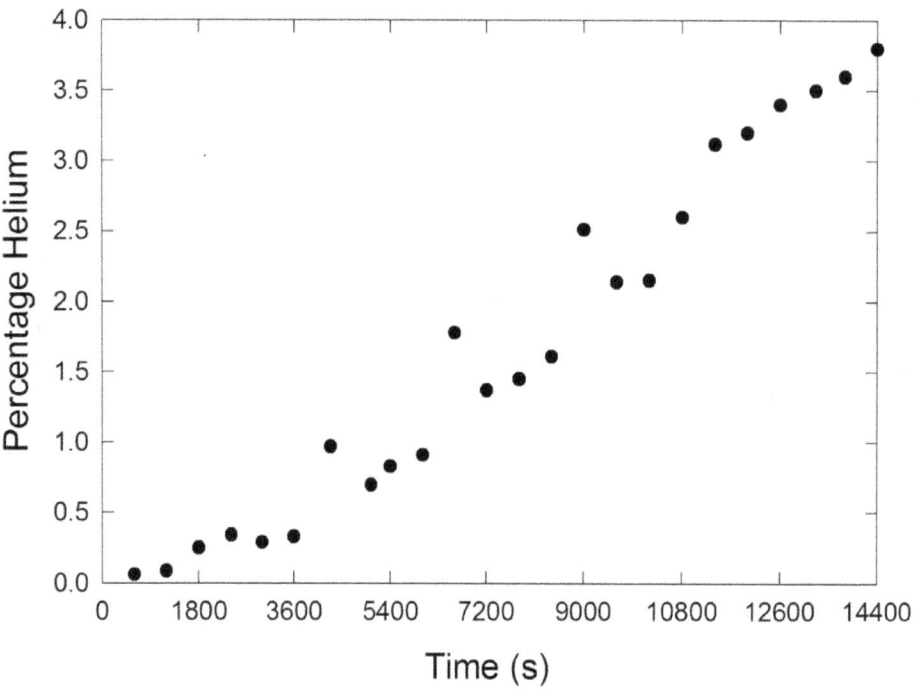

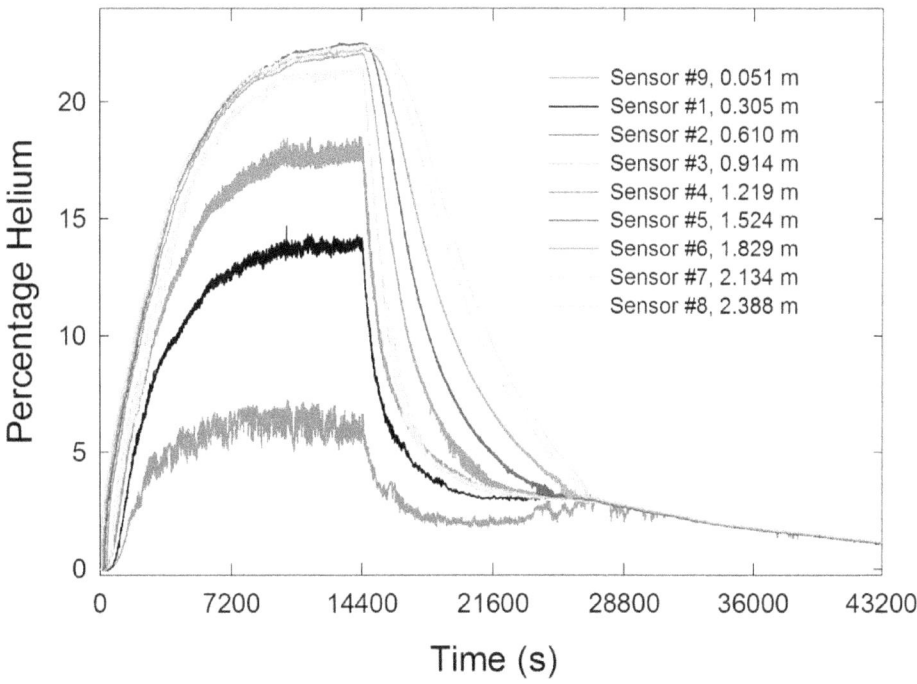

Figure 24. Helium volume percent is plotted as a function of time at the nine sensor heights indicated for the 7/29/10 experiment.

3.1.2 Helium release over large area without vehicle

Two helium release experiments for the empty garage similar to those performed in 2008 were done in 2010. The major change was that helium was released over a 0.0929 m^2 area as opposed to the 3.17×10^{-5} m^2 tube area of the earlier experiments. Helium volume flow rates were measured using the same approach. The time behaviors for the volume flow rates were similar to that shown in Figure 8. The average volume flow rates and total flow volumes are summarized in Table 4 for the experiments performed on 7/29/10 and 8/2/10.

Figure 24 shows plots of helium volume percent as a function of time for the nine sensor locations along the vertical array in the left rear quadrant. The overall time behaviors and magnitudes were similar to those observed in the 2008 experiments (see Figure 9 and Figure 18). Note that care should be employed when comparing plots since the vertical sensor array for the later experiment utilized different sensor separations and heights (see Table 1 and Table 2).

As observed in the earlier experiments, a uniform upper layer developed during the release. The quasi-steady-state helium volume fraction of 22.3 % listed in Table 5 was slightly less than observed in the 2008 experiments. Based on the variations of helium volume percent with height in Figure 24, it is clear that the interface separating the well-mixed upper layer from the lower volume where vertical concentration gradients developed was located just above 0.914 m. This is consistent with the 2008 results, which indicated that the interface location was between 0.76 m and 1.07 m above the floor.

The average helium volume percent in the garage at the end of the release was estimated using the approach described earlier. Figure 25 shows a plot of the quasi-steady-state concentrations versus height along with extrapolated values at the floor and ceiling and line approximations for the variations with height. Integration using the lines yielded an average value of 19.5 %. This value is higher than determined for the 2008 experiments even though the upper-layer quasi-steady-state concentration was lower. Comparison of Figure 25 with Figure 13 shows that for given heights in the lower layer the helium concentrations were higher for the 2010 experiment.

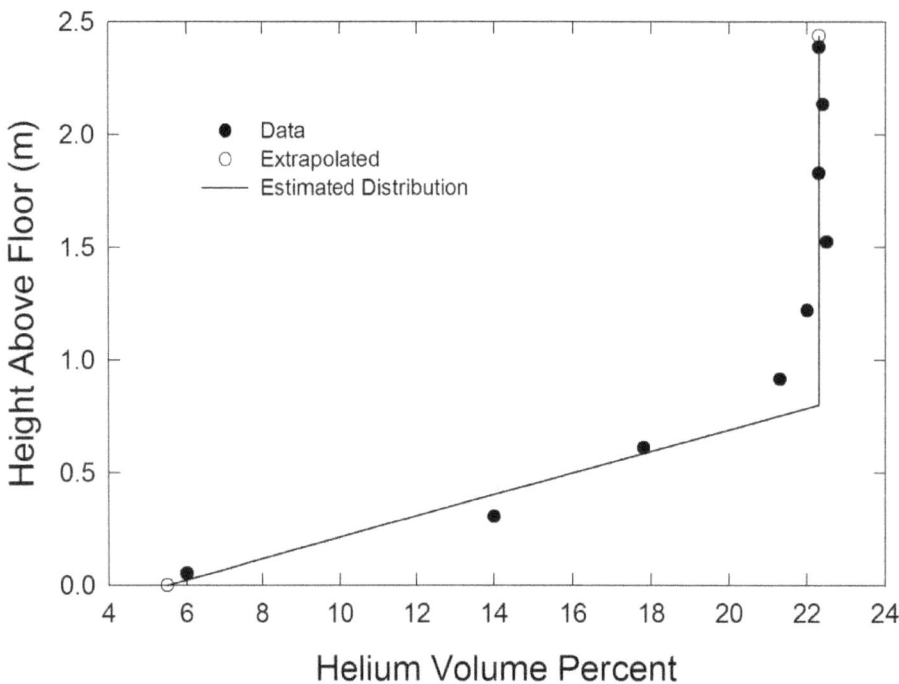

Figure 25. Experimental helium volume percents measured along the vertical array at the end of the 7/29/10 release are shown along with extrapolated values at the floor and ceiling. The solid lines are approximations of the data that are used to estimate the average helium concentration along the vertical direction.

The concentration profiles in the lower portion of the garage in Figure 24 show that the concentration gradient increased with decreasing height. The rapid and slow concentration fluctuations observed in the lower layer for the 2008 experiments were also present in the latest data. It is interesting that the quasi-steady-state helium volume percent was similar ($\approx$ 6 %) for measurements at 0.0152 m in 2008 and 0.051 m in 2010. This appears to have been the result of the different helium concentration distributions in the lower layer.

The helium concentration profiles following the end of the release also had similar appearances to the falloffs observed for the earlier results. One difference was that the helium volume percent at the lowest measurement height, which was nearer the floor than for the 2008 experiments, rose slightly as it approached the uniform helium concentration around 26 700 s. Values characterizing the falloff behavior of helium concentration during the post-release period are included in Table 6. The initial roll off of concentration at Sensor #8 was earlier than observed during the 2008 experiments. A shorter period was also required for the concentration at this height to collapse to the common decay curve. Even though these times were shorter, the helium volume percent for the collapsed curve at the time the falloff curve for Sensor #8 joined it was 3.0 %, which was intermediate between the two values from 2008. An exponential least squares fit to the common decay curve at long times yielded a time constant of 9.08×10^{-5} s^{-1} and ACH_{gar} = 0.33 h^{-1}. This value is somewhat higher than the values, 0.27 h^{-1} and 0.11 h^{-1}, observed for the experiments in 2008. The decay behaviors suggest that the gas exchange rate for the garage varied with time, with the 7/29/10 experiment being faster during the initial part of the post-release phase, but slower at longer times.

Additional helium sensors were added for the 2010 experiments. For the 7/29/10 experiment, four of these sensors were placed above the release location, and a fifth was added in the front-right quadrant at a height of 2.314 m. Figure 26 shows the helium volume percents recorded by these sensors. The large fluctuations observed for the sensors located over the helium release location show that the local

36

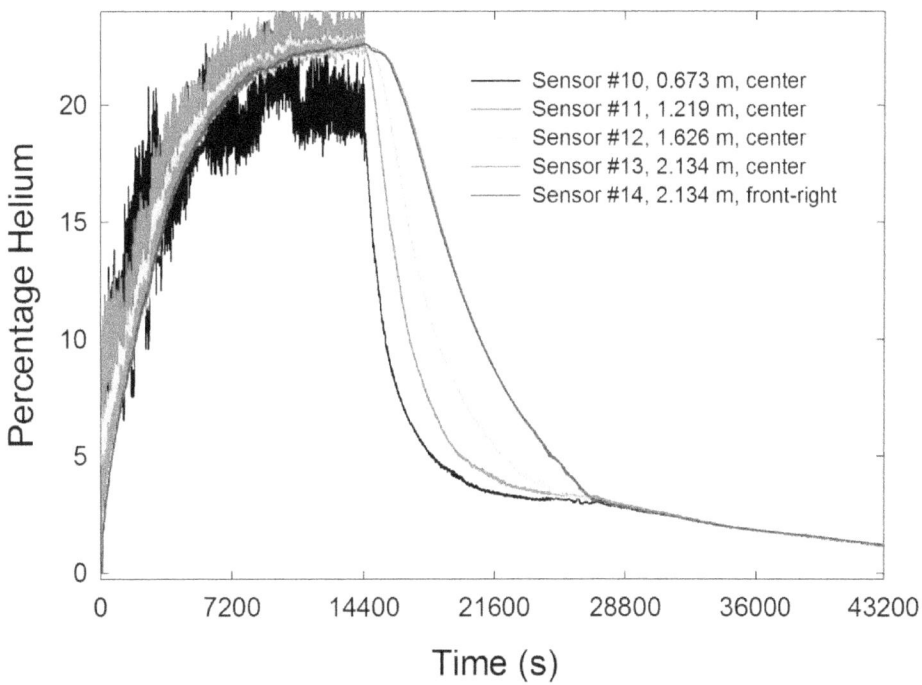

Figure 26. Helium volume percent is plotted as a function of time at the indicated heights for four sensors above the helium release location (center) and a single sensor at (x,y) = (4.343 m, 1.816 m) (front-right) for the 7/29/10 experiment.

concentrations fluctuated with time due to mixing inside the buoyant plume. The caveat concerning the effect of time-averaging on reducing the magnitudes of fluctuations noted earlier applies here as well. The concentrations at the various heights above the release location were higher, but not by large amounts, than those observed in the upper layer in the right front quadrant of the garage (Sensor #14).

The effects of mixing along the plume were particularly clear at early times, as can be seen in Figure 27, where the first three hundred seconds of the experiment are plotted. In less than a second after the flow started at 60 s, helium was detected at the lowest sensor 0.673 m above the release location. Helium was then sequentially observed at one second intervals at the three higher sensors above the release point. Somewhat surprisingly, the initial concentration was higher at the 1.219 m height than at the lower sensor. Normally, it would be expected that the helium concentration would decrease with distance from the release point. This observation may be tied to the mixing in the lower layer discussed above. An initial spike in helium concentration for the sensor located in the upper layer in the front-right quadrant was observed 15 s after the start of the release, and the concentration began to rise steadily 15 s later.

Over much of the release period shown in Figure 26, the helium volume percent at the lowest sensor position above the release location was the highest concentration, as expected. However, during much of the quasi-steady-state period, it was lower than observed for the higher sensors and even in the upper layer outside the plume. This is similar to the behavior seen at the start of the flow. All of the concentrations measured above the release locations were similar to the upper-layer value. Since the helium concentration leaving the distribution box should have been nearly 100 % helium (it is possible that some outside air mixed with the helium by back flow before it exited the box), this implies that a great deal of mixing with surrounding air took place between the release height and the lowest sensor 0.673 m above the floor.

As soon as the helium flow stopped, the rapid fluctuations dissipated, and helium concentrations for the three highest sensors (1.219 m and higher) above the release location and in the front-right upper layer collapsed to a common helium volume percent of 22.5 %. This is very close to the upper-layer value of

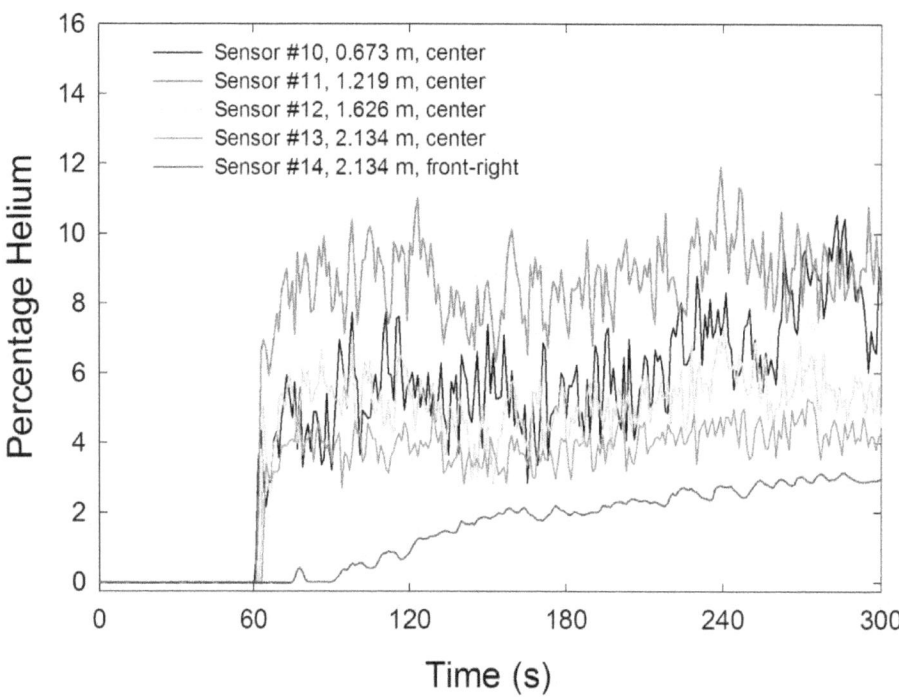

Figure 27. Helium volume percent is plotted as a function of time for the five sensor heights indicated for the initial 300 s of the 7/29/10 experiment.

22.3 % given in Table 5 based on measurements along the vertical sensor array in the left-rear of the garage. The corresponding helium concentration for the lowest sensor at 0.673 m was estimated to be 19.8 %, which falls midway between the corresponding concentrations of 17.8 % and 21.3 % recorded by the vertical-array sensors at heights of 0.610 m and 0.914 m, respectively.

The initial falloff behaviors following the end of the release for the data in Figure 26 were similar to those described for the sensors along the vertical array, with the drops in concentration starting immediately for the lowest sensor and progressing with time upwards through the uniform upper layer. The concentration falloffs collapsed to a common helium volume percent around 26 700 s or 12 300 s after the end of the release. This compares favorably with the value of 12 400 s included in Table 6 based on the measurements along the vertical array. The helium volume percents when the concentration collapses occurred were roughly 3.1 % and 3.2 % for the sensors in the rear-left quadrant and above the release location, respectively. During the post-release phase, it is difficult to identify the data for the sensor located at a height of 2.134 m above the release location in Figure 26 because the curve for the sensor located at the same height in the right-front of the garage overlays it. Comparison with the response for Sensor #7 shown in Figure 24, which was located at the same height in the left-rear quadrant, shows that its falloff was indistinguishable from those at the center and right front. This provides strong evidence that there were no horizontal concentration gradients at this height, and the upper layer was fully mixed.

Neither wind speed nor differential pressure measurements were recorded for the 7/29/10 experiment.

Figure 28 compares temperatures measured in the family room of the house, inside the garage, and outside. The outside and family room temperatures were taken from the measurement system for the house, and the garage temperatures were recorded manually using a thermocouple reader. It was a hot summer day, and temperatures in the garage and outside were 8 °C to 11 °C higher than the air-conditioned interior. Temperature fluctuations in the family room reflect the cycling of the air conditioning. The fan for the air conditioning was operated continuously.

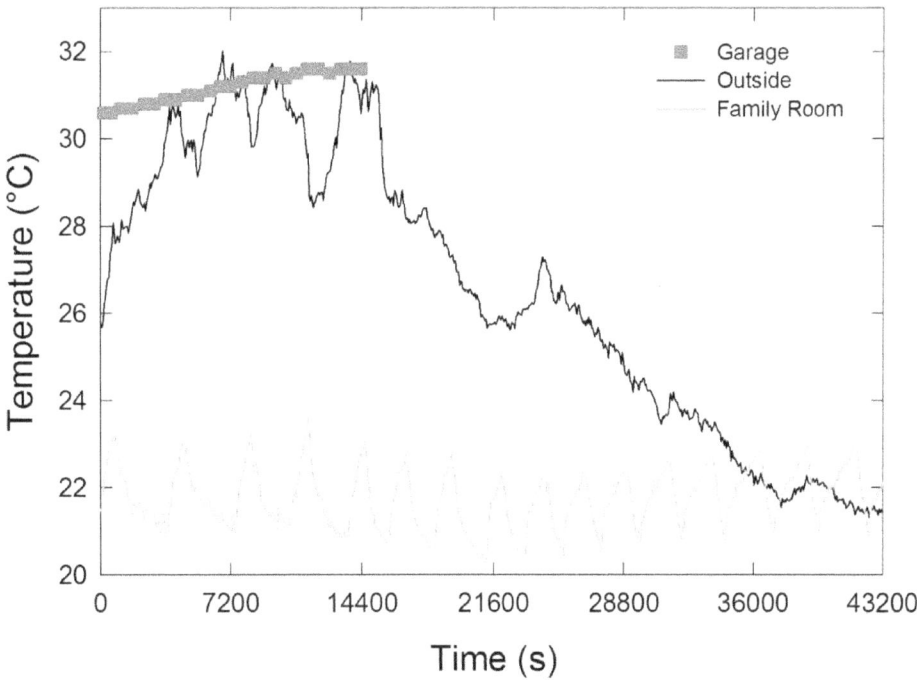

Figure 28. Temperatures measured using thermocouples (symbols) and thermistors (lines) are plotted as a function of time for measurements recorded on 7/29/10-7/30/10 in the family room, garage, and outside.

Figure 29 shows a plot of the helium volume percent measured in the family room over a period of twenty-four hours, including the 4 h release period. Measurements recorded manually and electronically agree well. Once the helium release began at 60 s, the helium volume percent in the house rose continuously, reaching a maximum value of 3.7 % approximately 400 s after the end of the release. This value is slightly less than that observed in the 9/12/08 experiment (see Figure 23.). Corresponding values of helium volume and fraction of released helium in the house at the end of the release are listed in Table 7. Very shortly after the end of the release, the helium concentration began to fall slowly, approaching zero after 20 h. Some fluctuations are evident during the decay phase, which appear to be correlated with the air conditioning cycling. The source of the fluctuations is unclear.

The helium release into the empty garage was repeated on 8/2/10. The average helium volume flow rate and total flow volume are included in Table 4. Figure 30 is a plot of the helium volume percent for the nine locations along the vertical sensor array in the left-rear quadrant of the garage. Both the general appearance and quantitative behavior are similar to the results for 7/29/10 shown in Figure 24. The most obvious differences between the two experiments occurred during the post-release period. The helium volume percent recorded by the lowest sensor in the second experiment did not show the minimum that was evident on 7/29/10 as the concentrations collapsed to a common curve. Careful comparison showed that the collapsed curve for the later experiment fell slightly below that for the earlier. These observations suggest a slightly faster helium loss rate for the repeated experiment.

Quantitative characterizations of the data in Figure 30 are included in Table 5 and Table 6 for the release and post-release phases. The quasi-steady-state upper layer and average concentrations at the end of the release were both slightly higher than observed on 7/29/10, as was the volume of helium remaining in the garage at this time. The characteristics for the post-release period in Table 6 are similar to those determined from earlier measurements. The largest difference was for the helium volume percent at the time when the decay curves collapsed to a single curve, which at 2.2 % was the lowest of the experiments discussed thus far. This was the case even though the time required to decay to this level was

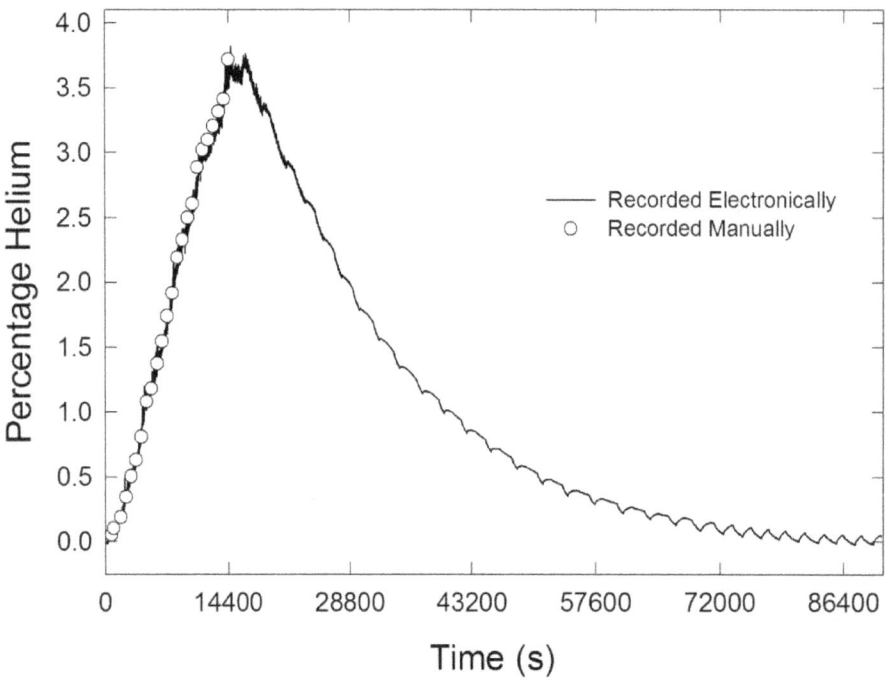

Figure 29. The helium volume percents measured in the family room of the house manually (symbols) and electronically (line) are shown as a function of time for the 7/29/10 experiment.

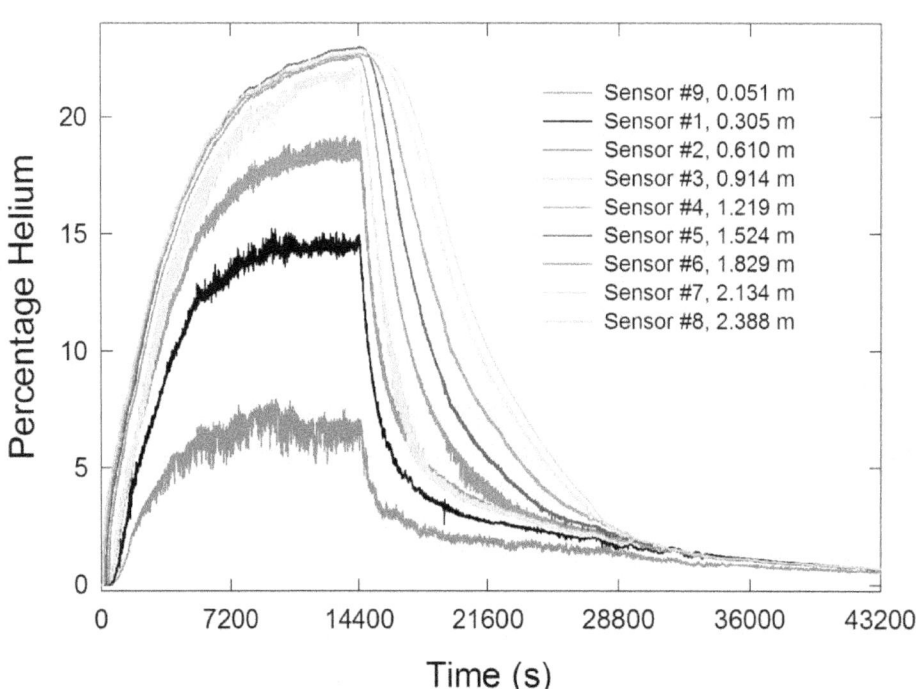

Figure 30. Helium volume percent is plotted as a function of time at the nine sensor heights indicated for the 8/2/10 experiment.

intermediate to the other experiments. The $ACH_{gar} = 0.22$ h^{-1} determined at long times was also near the median for the earlier experiments.

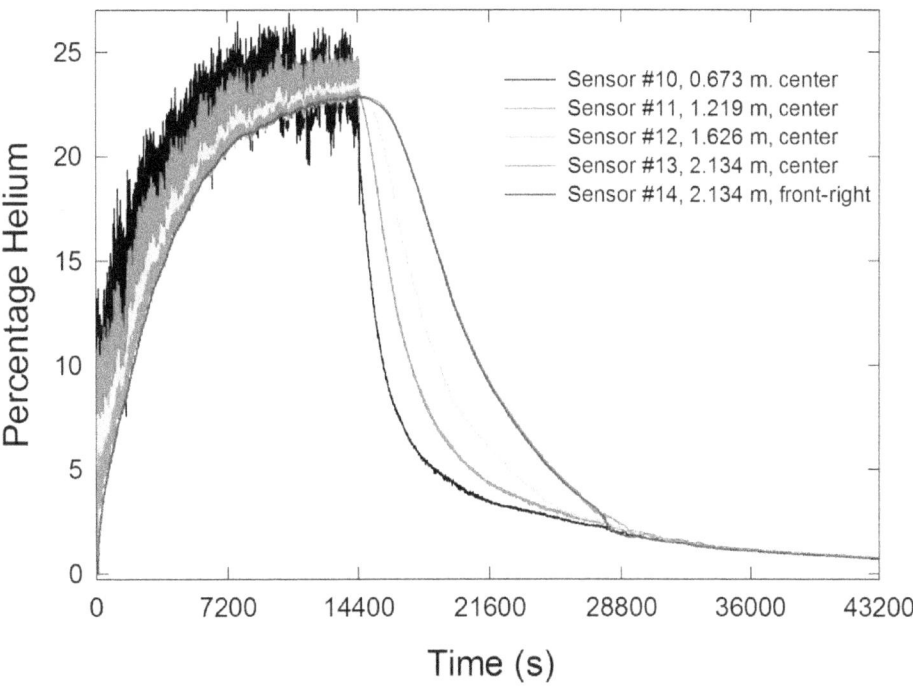

Figure 31. Helium volume percent is plotted as a function of time at the indicated heights for four sensors above the helium release location (center) and a single sensor at (x,y) = (4.343 m, 1.816 m) (front-right) for the 8/2/10 experiment.

The concentration time profiles recorded by the remaining five sensors for the 8/2/10 experiment are shown in Figure 31. The corresponding results for 7/29/10 were plotted in Figure 26. Comparison shows that concentrations along the centerline above the release location were slightly higher ($\approx$ 5 %) during the second experiment. This difference was several times larger than the relative difference in helium volume flow rates listed in Table 4. There were also differences in concentration for Sensor #10, centered 0.673 m above the helium release location, near the end of the release. For the 7/29/10 data, the helium concentrations were strongly fluctuating and were apparently lower than recorded by sensors located above the release point. Such behavior was not as evident for the second run. A similar difference is also evident in the plot of the data over the initial 300 s of the experiment shown in Figure 32. The concentrations show the expected falloff from top to bottom for sensor locations above the release location. Unlike for the 7/29/10 results, the helium volume percent for Sensor #10 abruptly dropped to 19.6 %, very close to the value of 19.8 % estimated from the 7/29/10 experiment, when the helium flow was stopped and then began to fall at a slower rate. These observations suggest that conditions in the lower layer had a reduced influence on the helium plume during the second experiment.

When the helium flow was halted, the fluctuations at all four locations above the plume rapidly disappeared, and the smooth falloff curves were similar to those observed on 7/29/10. As before, the two curves for a height of 2.134 m (above the release point and in the front right) were superimposed. Similar to the results in the left rear, the concentration falloffs eventually collapsed to a single curve.

The differential pressure between the interior of the garage and outside was recorded on 8/2/10. The results are plotted in Figure 33. Note that the averaging time for the measurements was changed from 1 s to 10 s at 28 800 s, which is the reason for the abrupt decrease in fluctuation magnitude. The differential pressure behavior was similar to those observed on 9/11/08 (Figure 12) and 9/12/08 (Figure 19) and suggests that wind levels were not dramatically differently for these three experiments.

Figure 34 shows plots of temperature for the 12 h time period following the start of the experiment on 8/2/10 for locations in the family room, garage, and outside. The day was cloudy and breezy, and the temperatures in the garage and outside were somewhat lower than measured on 7/29/10. The fluctuations

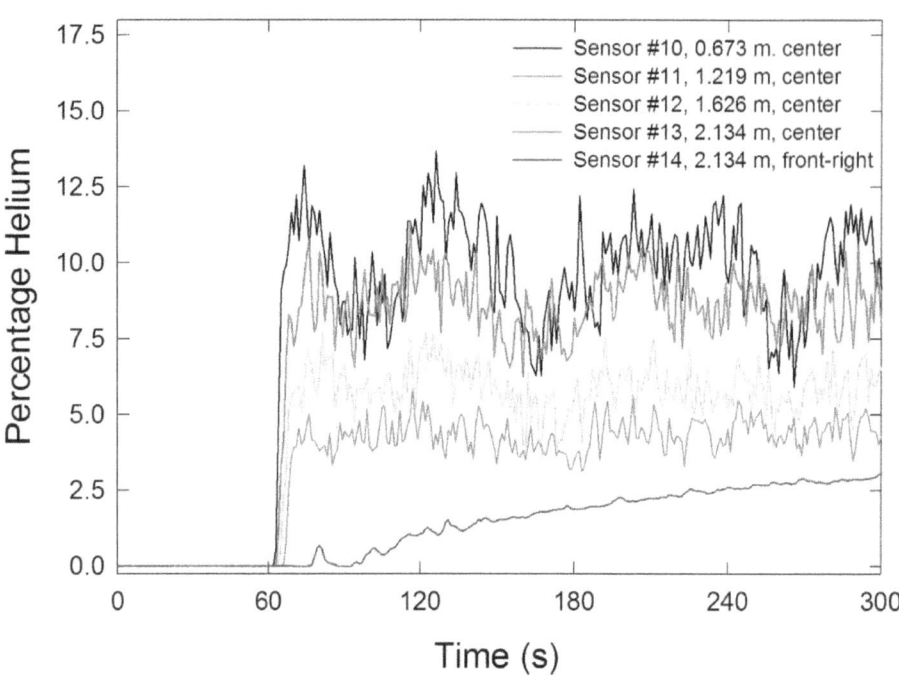

Figure 32. Helium volume percent is plotted as a function of time for the five sensor heights indicated for the initial 300 s of the 8/2/10 experiment.

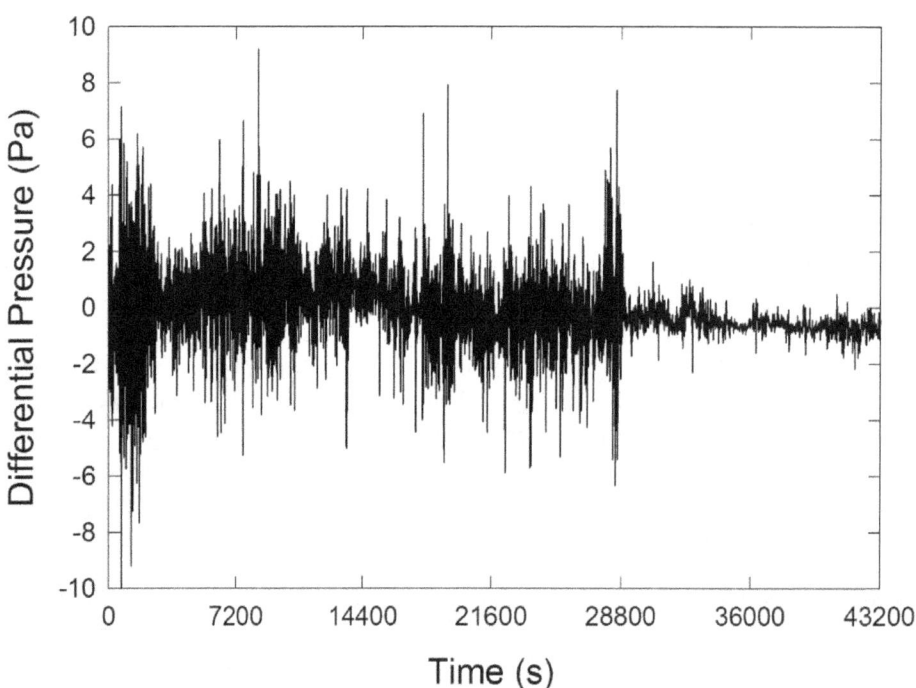

Figure 33. The time variation of the differential pressure between the garage interior and outside is plotted over the time period from 0 s to 43 200 s for the 8/2/10 experiment.

due to the cycling of the air conditioning were evident in the family room. The blower for the air conditioning was operated continuously.

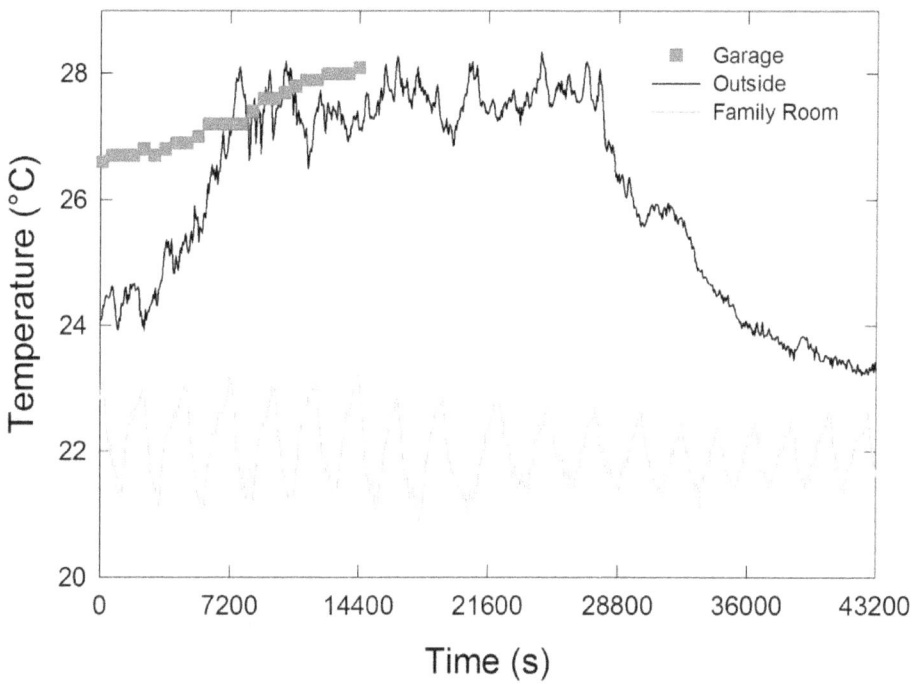

Figure 34. Temperatures measured using thermocouples (symbols) and thermistors (lines) are plotted as a function of time for measurements recorded on 8/2/10 in the family room, garage, and outside.

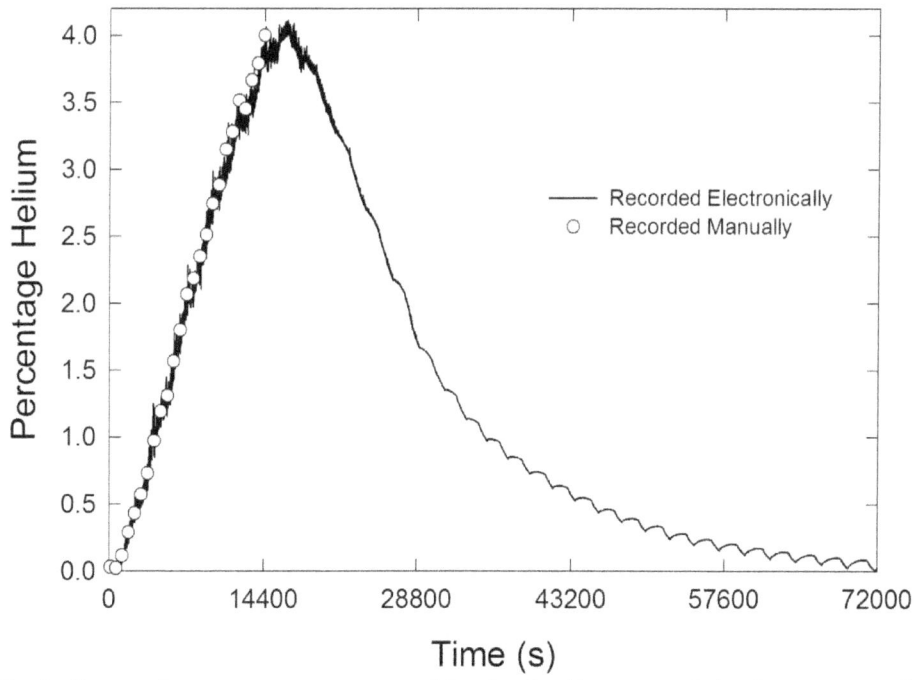

Figure 35. The helium volume percents measured in the family room of the house manually (symbols) and electronically (line) are shown as a function of time for the 8/2/10 experiment.

The helium volume percent measured in the family room for the 8/2/10 experiment is shown in Figure 35. The growth and decay of the helium concentration were similar to those for the 7/29/10 experiment shown in Figure 29, but the maximum concentration for the later experiment is 4.1 % versus

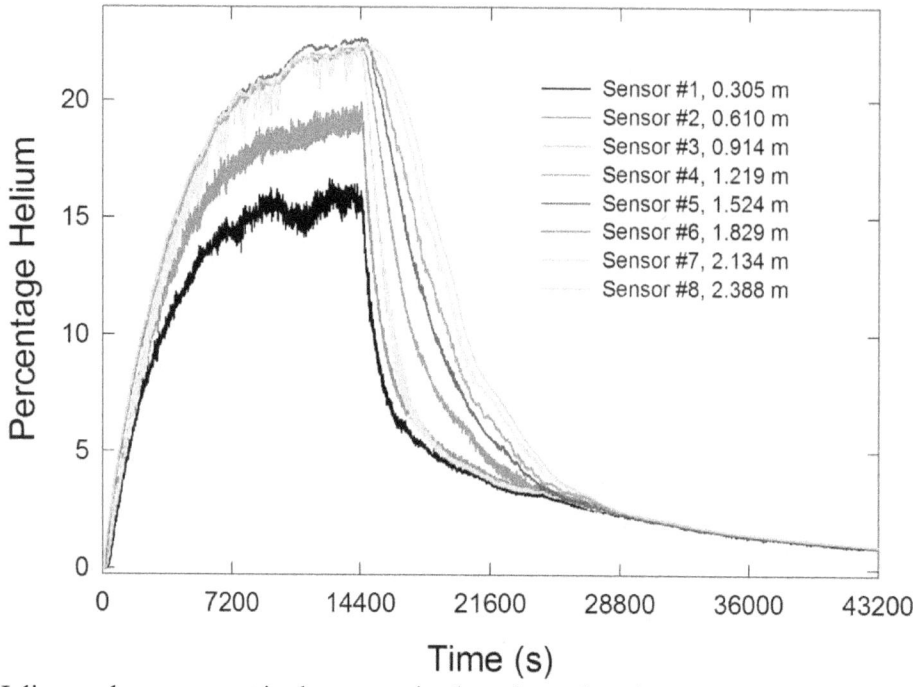

Figure 36. Helium volume percent in the garage is plotted as a function of time at the eight sensor heights indicated along the vertical array for the 8/6/10 experiment.

3.7 % for the earlier. This difference is larger than the difference in volume flow rates and suggests that a higher fraction of helium was transported into the house from the garage during this experiment. This is confirmed by comparing the values of fraction of released helium in the house listed in Table 7. The maximum value was recorded 450 s after the end of the helium release inside the garage. Fluctuations due to the cycling of the air conditioning are apparent. The source of these fluctuations was not clear.

3.1.3 Helium release over large area with vehicle

Two 4 h releases of helium were made with an automobile parked at the center of the garage. Helium from the release box flowed upward and struck the undercarriage of the automobile before spreading out under the automobile, accumulating at various locations, and eventually escaping to the surrounding garage.

The first experiment was run on 8/6/10. The measured volume flow parameters are included in Table 4. The helium volume flow rate and total volume released were slightly higher than for the earlier experiments. The helium flow was initiated at 60 s and lasted 14 400 s.

The locations of the sensors were changed from those used in the experiments without an automobile. Sensor #9 was moved from its position near the floor on the vertical sensor array to the front driver's side wheel well and Sensors #10 to #14 were placed at other locations within the vehicle.

The concentration time behaviors recorded by the vertical array of sensors are shown in Figure 36. Comparison with the corresponding plots (Figure 9, Figure 18, Figure 24, and Figure 30) recorded without a vehicle shows that the growth of the concentration profiles during the helium release phase and the decay during the post-release phase had the same general appearances. The helium volume percents grew rapidly during the initial release phase before leveling off, while the concentrations dropped following the end of the release, with the concentrations at the lowest sensor falling the earliest and fastest and concentrations at higher positions decaying later. The helium concentrations for all sensors ultimately collapsed to a single nonzero value, which then decayed slowly with time.

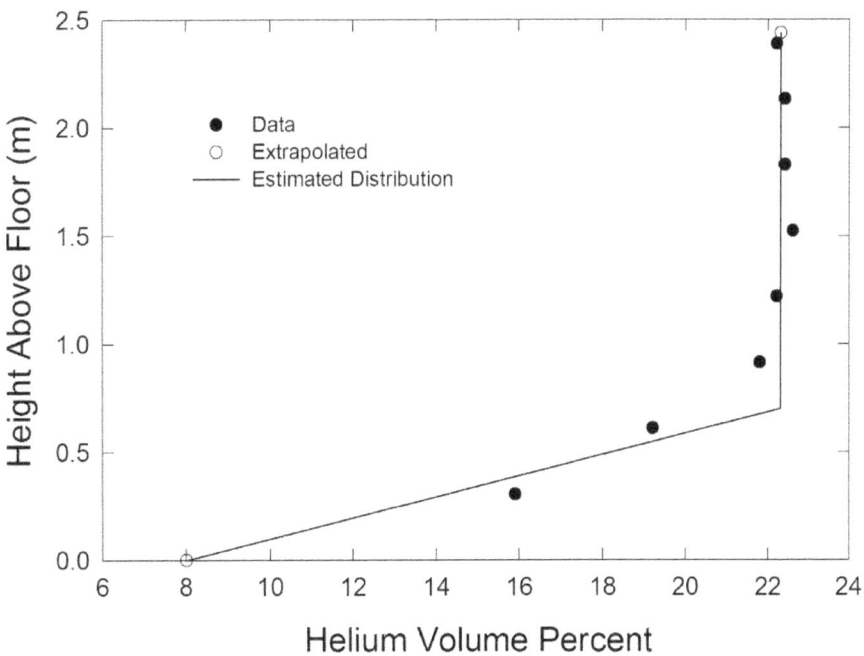

Figure 37. Experimental helium volume percents measured along the vertical array at the end of the 8/6/10 release are shown along with extrapolated values at the floor and ceiling. The solid lines are approximations of the data used to estimate the average helium concentration along the vertical direction.

Table 5 lists values of the parameters used to characterize the experiment at the end of the release period. These are nearly identical to those determined for the 8/2/10 data, but the two sets of results differ somewhat when compared to the earlier experiments without a vehicle. There is evidence that mixing inside the garage was improved slightly by the presence of an automobile. As observed in the experiments without a vehicle (compare Figure 36 with Figure 24 and Figure 30), there was a well-mixed region in the upper part of the garage. However, the location of the interface with a vehicle present was slightly lower with Sensor #3 at 0.914 m reaching a quasi-steady-state concentration very close to that characteristic of the upper layer, even though sharp concentration spikes were evident at this height. For the cases without a vehicle on 7/29/10 and 8/2/10, the quasi-steady-state helium volume percents at this height were clearly lower than in the well-mixed upper layer. Additional evidence for increased mixing in the presence of the vehicle can be seen in Figure 37, which shows the quasi-steady-state helium volume percents as a function of height along the sensor array. Quasi-steady-state helium concentrations in the lower layer at 0.305 m (Sensor #1) and 0.610 m (Sensor #2) were considerably higher with the automobile present. As a result, comparison with Figure 13 and Figure 25 shows that the average concentration gradient in the lower layer was reduced compared to experiments without a vehicle present.

The decay of the helium volume fraction in the garage following the end of the release for the 8/6/10 experiment also appeared to be very similar to the corresponding results for 8/2/10 in Figure 30. This is confirmed by the agreement between quantitative characteristics for the two experiments included in Table 6. The close agreement extends to parameters that characterize the concentration falloff immediately following the end of the helium release, as well as the decay constants and ACH_{gar} derived from measurements at longer times when helium levels in the garage were 1 % and lower.

Figure 38 shows differential pressure measurements between the garage and the outside. During the early part of the helium release, there were negative pressure fluctuations that were generally more intense and larger than were observed in earlier experiments (see Figure 12, Figure 19, and Figure 33).

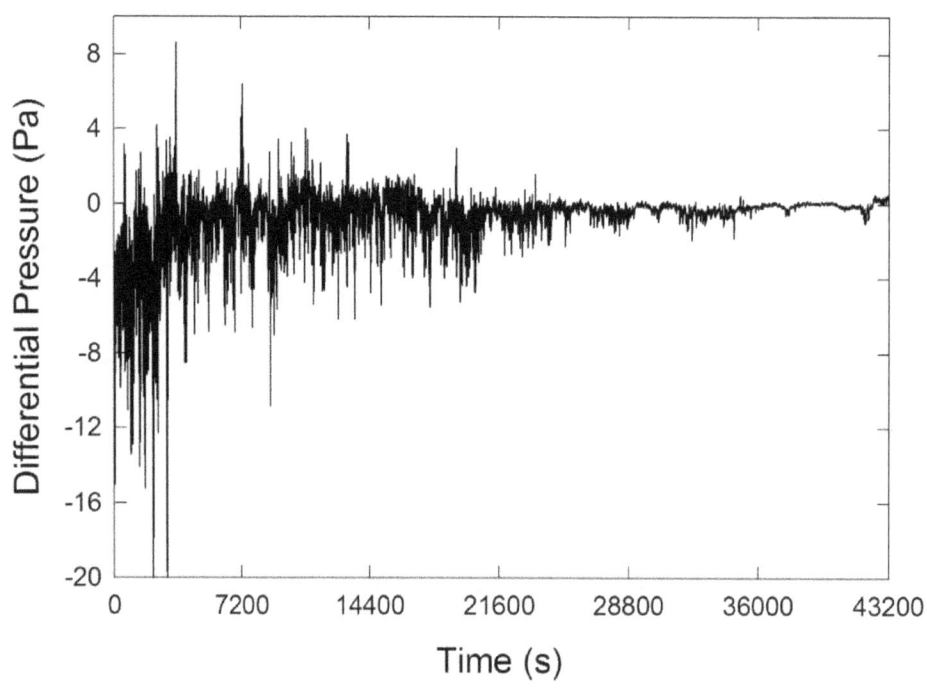

Figure 38. The time variation of the differential pressure between the garage interior and outside is plotted over the time period from 0 s to 43 200 s for the 8/6/10 experiment.

Even though the fluctuations were quite large, the effects on the loss rates of helium from the garage were relatively small. Note that the data averaging time was changed from 1 s to 10 s at 28 800 s.

The temperatures recorded in the family room, garage, and outside during the 8/6/10 experiment are shown in Figure 39. It was a hot summer day, and during the helium release the outside and garage temperatures were roughly 6 °C higher than in the air conditioned house. Shortly after the end of the helium release, the compressor on the air conditioning system failed, and the interior temperature began to increase. At some point shortly afterwards, windows on the front and rear of the house were opened.

A plot of helium volume percent recorded in the family room during the 8/6/10 experiment is shown in Figure 40. Table 7 summarizes parameters related to the buildup of helium in the house. The growth curve of the helium concentration was similar to that observed in earlier experiments, but the maximum level reached, 3.2 %, was somewhat lower. The reduced concentration was likely due to a higher exchange rate with the outside air, which is consistent with the wind observations, or to a reduced gas exchange rate between the garage and house, which might have been due to the relatively large temperature difference between the two locations. During the post-release phase, the falloff of the family room helium concentration was much faster than observed when the vehicle was not present (compare Figure 40 with Figure 29 and Figure 35). The most likely reason was the opening of the house windows, even though the exact time when this occurred was unknown.

During the 8/6/10 experiment, helium concentrations were measured in the undercarriage of the vehicle near the helium release location, in the passenger-side front wheel well, at two locations within the engine compartment, and inside the passenger compartment and trunk. Figure 41 shows the time behaviors for the helium volume percent at these locations during the release and post-release periods. Note that the thermal conductivity sensor (Sensor #10) inside the passenger compartment was originally attached to the compartment light near the center of the roof, but that it fell at some unknown time and was found on the floor behind the passenger seat at the end of the experiment.

The data shown in Figure 41 are replotted in Figure 42 on an expanded time scale to emphasize the concentration behavior during the initial part of the helium release. It is evident that the helium concentration in the volume underneath the vehicle rose sharply very shortly (within 4 s) after the helium

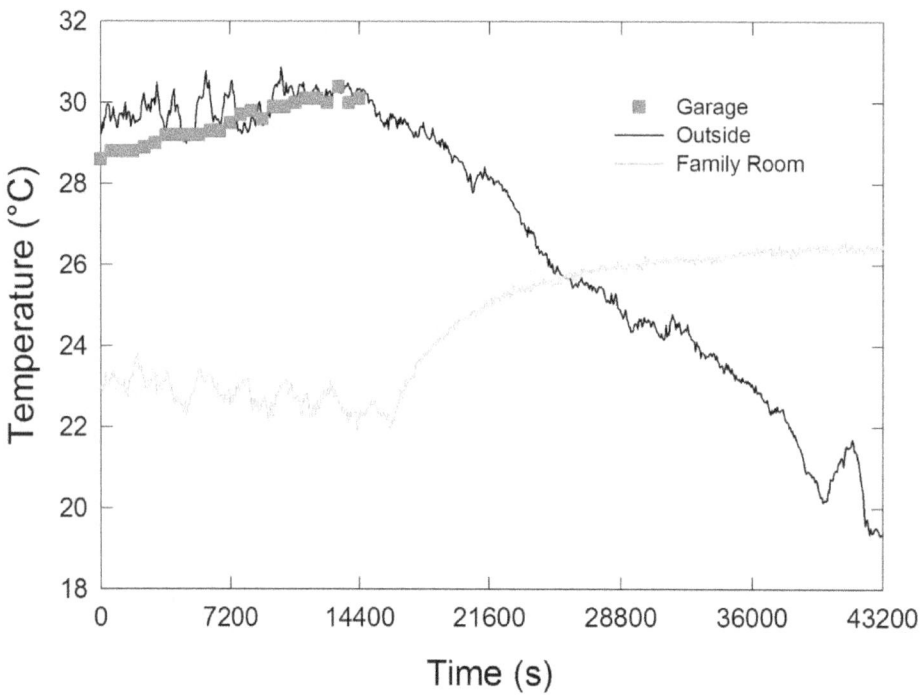

Figure 39. Temperatures measured using thermocouples (symbols) and thermistors (lines) are plotted as a function of time for measurements recorded on 8/6/10 in the family room, garage, and outside.

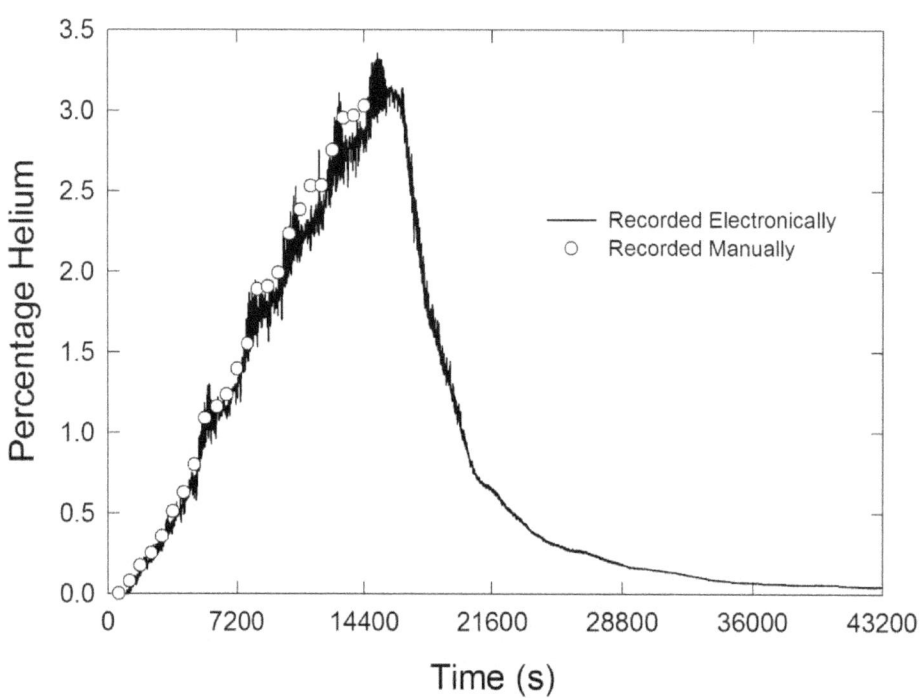

Figure 40. The helium volume percents measured in the family room of the house manually (symbols) and electronically (line) are shown as a function of time for the 8/6/10 experiment.

flow was started at 60 s and rapidly attained a helium concentration of roughly 28 %. This concentration was close to the stoichiometric concentration for hydrogen burning in air [33], which indicates that the

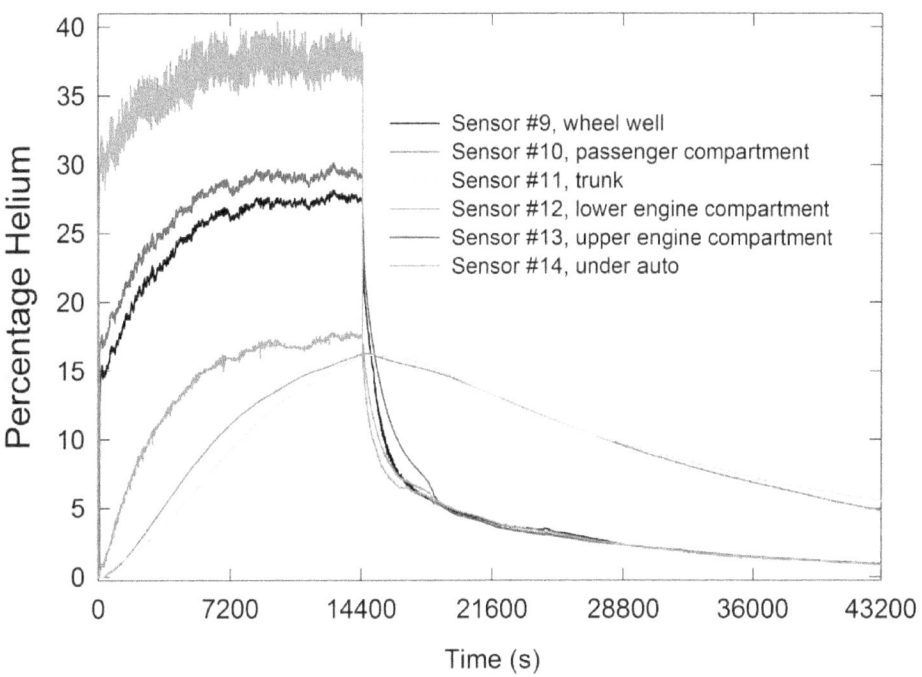

Figure 41. Helium volume percent is plotted as a function of time at the indicated locations inside the automobile centered over the helium release location for the 8/6/10 experiment.

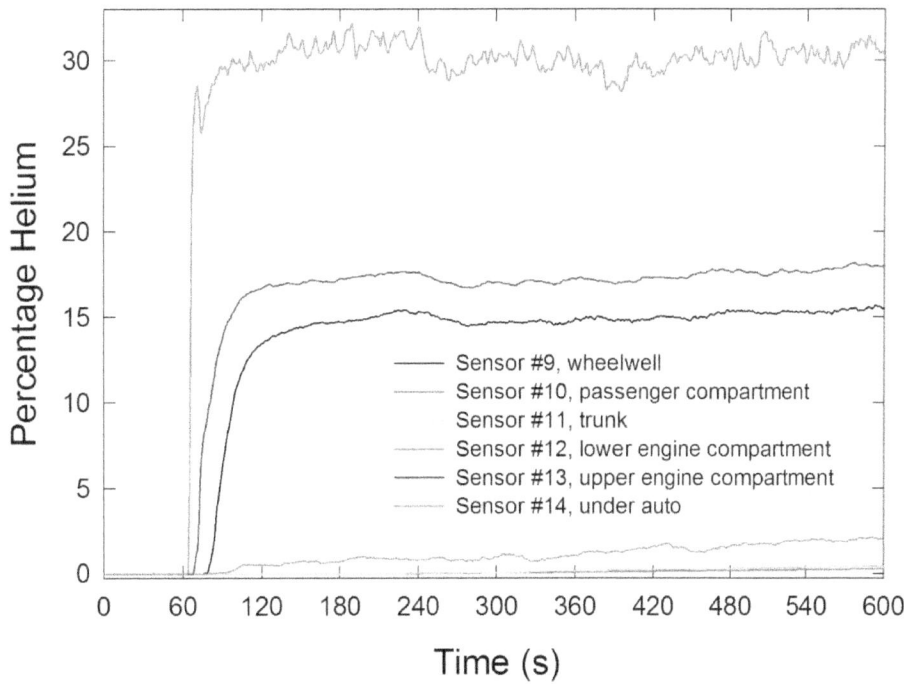

Figure 42. The data shown in Figure 41 are replotted on an expanded time scale emphasizing the helium increases when the flow was initially released.

experimental configuration used here would have rapidly generated a dangerous flammable mixture at this location if hydrogen had been released instead of helium. It can be seen in Figure 41 that after

49

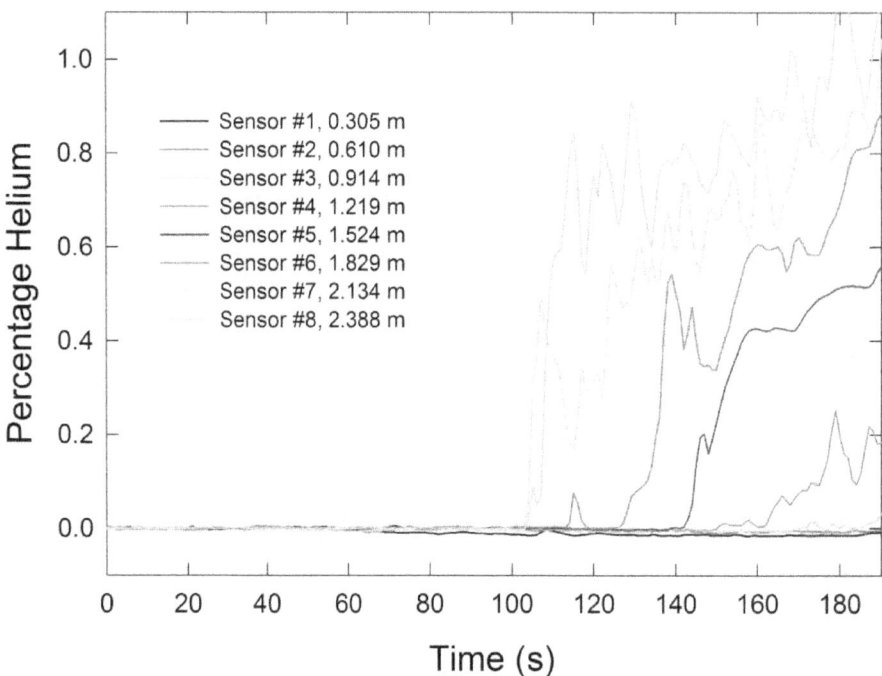

Figure 43. Helium volume percent is plotted as a function of time for the eight sensor heights indicated for the initial 190 s of the 8/6/10 experiment.

reaching this level the concentration slowly increased further until it reached a quasi-steady-state value around 37 % prior to the end of the release.

The corresponding plot for the initial part of the release period for the eight sensors along the vertical array is shown in Figure 43, which was obtained by expanding the time scale for the data in Figure 36. It can be seen that 43 s was required for the initial detection of helium at the 2.39 m height of Sensor #8, which is nearly 40 s after helium was first detected under the automobile. This delay can be compared to a value of 15 s measured for the corresponding data shown in Figure 30 for a case without a vehicle present. This indicates that when releasing the helium under the vehicle, an additional period of 25 s was required for helium to reach a sensor location close to the ceiling.

The high concentration of helium that built up within the undercarriage of the Stratus apparently flowed to other accessible areas in the vehicle. The next location where helium was detected following the start of the release was at the top of the engine compartment about 4 s later. The helium volume percent increased rapidly to about 17 %. The nearly factor of two drop in concentration between the undercarriage measurement location and the top of the engine compartment shows that additional air was mixing into the flow either by entrainment from below or through openings in the engine compartment. The next sensor reached by the helium was located in the wheel well 0.76 m above the floor. The concentration increased in a manner similar to that at the top of the engine compartment and reached a level that was only slightly lower. Since a pathway existed between the engine compartment and the wheel well, this suggests that a relatively uniform concentration of helium developed in the upper part of the engine compartment.

The area of relatively high concentration did not extend uniformly over the full height of the engine compartment. While the helium concentration at the front of the engine 0.41 m above the floor began to increase shortly after the other sensors detected helium, the rate of increase was more gradual, and the quasi-steady-state levels attained were only slightly more than half of those in the wheel well and upper portion of the engine compartment. This indicates that there was a significant falloff in helium concentration in the lower part of the engine compartment. In fact, the quasi-steady-state helium volume percent of around 17.5 % is very close to the extrapolated concentration at the same height measured

50

along the vertical array (see Figure 36). This suggests that very little of the higher concentration helium/air mixture trapped underneath the vehicle reached this location.

The helium concentrations trapped in the undercarriage and engine compartment of the automobile during the release phase indicate that hydrogen released at similar flow rates would create significant volumes of flammable hydrogen/air mixtures a few seconds following the start of a release. Due to the higher concentrations trapped within the vehicle as compared to the garage interior, the combustion process inside these enclosed spaces would be expected to be more vigorous since the hydrogen/air flame speed increases until the concentration exceeds the stoichiometric ratio. [33]

During the helium release, concentrations underneath the vehicle and within the engine compartment, after jumping abruptly at the start of the release, grew slowly with time before attaining quasi-steady-state levels. The general time behaviors were similar to those observed for the sensors located along the vertical array inside the garage. This shows that the helium concentrations within the vehicle grew due to mixing of the released helium with surrounding gases in the garage for which the helium concentration was increasing. The quasi-steady-state inside the vehicle only developed when it developed in the garage.

The growing helium concentrations underneath the automobile and in the engine compartment as well as in the surrounding garage penetrated the passenger compartment and trunk of the car as is evident in Figure 41. However, the exchange rate was far from instantaneous. The roughly comparable helium concentrations at these two locations increased at considerably slower rates than elsewhere under the vehicle and outside and were still increasing at the end of the 4 h release period, despite the development of quasi-steady-state concentrations elsewhere. This shows that the passenger compartment and trunk behaved as partially enclosed spaces with relatively lower exchange rates with their surroundings than the garage, which attained a quasi-steady-state helium distribution in about two hours. Assuming hydrogen penetration would behave in a similar way, the results indicate that the development of flammable mixtures inside the passenger compartment and trunk would require longer periods than in the other partially enclosed spaces of the vehicle or inside the garage.

When the helium flow was shut off, the high helium concentrations underneath the vehicle and in the engine compartment rapidly dissipated to levels comparable to those of the lower layer inside the garage (compare Figure 41 and Figure 36). The most rapid drop was for the location under the vehicle (Sensor #14). By expanding the curve in Figure 41, it was found that the helium volume percent at this location dropped by half of the difference between the level present at the end of the release and the value for Sensor #1 in roughly 7 s. The falloff rate decreased somewhat after this, but the concentration still approached the external value in 30 s. The falloffs at the tops of the engine compartment and wheel well were somewhat slower. They initially fell to levels that were intermediate between those recorded by Sensors #2 and #3, and then tracked the falloff of these concentrations. Roughly 50 s was required for the concentrations to decrease by one half of the difference between the engine compartment and the garage interior at comparable heights. The sensor located in the lower front of the engine compartment measured helium concentrations that were a couple of percent higher than recorded by the lowest sensor in the garage (Sensor #1, 0.305 m above floor). This remained the case after the helium flow was stopped. This provides further evidence that very little of the helium trapped under the vehicle reached this location.

The helium concentration trapped inside the passenger compartment and trunk decayed much more slowly after the helium flow was stopped than observed for other locations in the vehicle or the garage. This is evident in Figure 41 where the helium volume percent levels inside the passenger compartment and trunk rapidly became higher than those trapped under the car or in the engine compartment once the helium flow was stopped. The falloff rates of the helium concentration inside the passenger compartment and trunk were clearly slower than observed elsewhere in the vehicle. Helium concentrations were monitored at the locations in the garage and vehicle until 240 000 s following the start of the experiment. Inspection revealed that concentrations outside of the passenger compartment and trunk tracked each other, eventually falling to zero around 85 000 s after the end of the release. Concentrations in the passenger compartment and trunk, which tracked each other closely, required over 200 000 s to reach this level. These observations show that there was rapid gas exchange between the passenger compartment

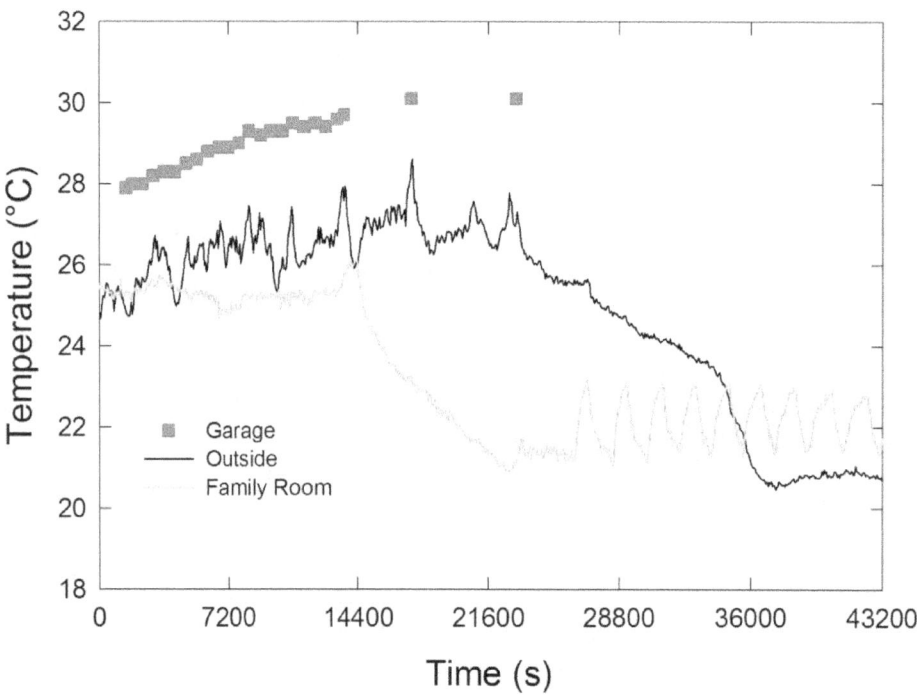

Figure 44. Temperatures measured using thermocouples (symbols) and thermistors (lines) are plotted as a function of time for measurements recorded on 8/23/10 in the family room, garage, and outside.

and trunk, but the exchange between these volumes and their surroundings was much slower than between the garage interior and its surroundings.

The repeat experiment with a vehicle in the garage was run on 8/23/10. The weather was relatively cool and sunny. The central air conditioning for the house was still not functioning at the start of the experiment. A small window unit air conditioner located in the family room was used to maintain a nearly constant temperature of 25 °C in the family room area where the instrumentation was located. At approximately 13 200 s after the start of the experiment, repairs to the central air conditioning were completed, and the system began to cool the entire interior of the house.

Figure 44 shows temperatures recorded in the family room, garage, and outside during the experiment. The temperature in the family room actually increased slightly when the central air conditioning was initially turned on. This was likely due to recirculation of higher temperature air from elsewhere inside the house. Shortly afterwards, the temperature began to fall from 26 C to 21 °C over a period of 8 500 s. After this time the air conditioning began to cycle in the normal manner. Outside temperatures were similar to those inside the house, while temperatures in the garage were a couple of degrees higher.

The volume flow rate for helium had the same type of oscillations evident in Figure 8. The measured helium volume flow rate, flow time, and total volume release are included in Table 4. The helium supply was exhausted before the 4 h release period was complete, and, as a result, the total volume of helium released was slightly less than for the 8/6/10 experiment with a vehicle (59.1 m^3 versus 64.3 m^3).

Figure 45 and Figure 46 show time behaviors of helium volume percent measured for the various heights along the vertical array in the garage and for the six sensors located inside the vehicle, respectively. Comparison of Figure 45 and Figure 36 reveals that the vertical profiles for the two experiments with vehicles were similar when the difference in helium release periods is accounted for. This was the case for the initial growth and the approach to a quasi-steady state during the release phase, as well as the concentration decays during the post-release phase. The close agreement in Table 5 and

52

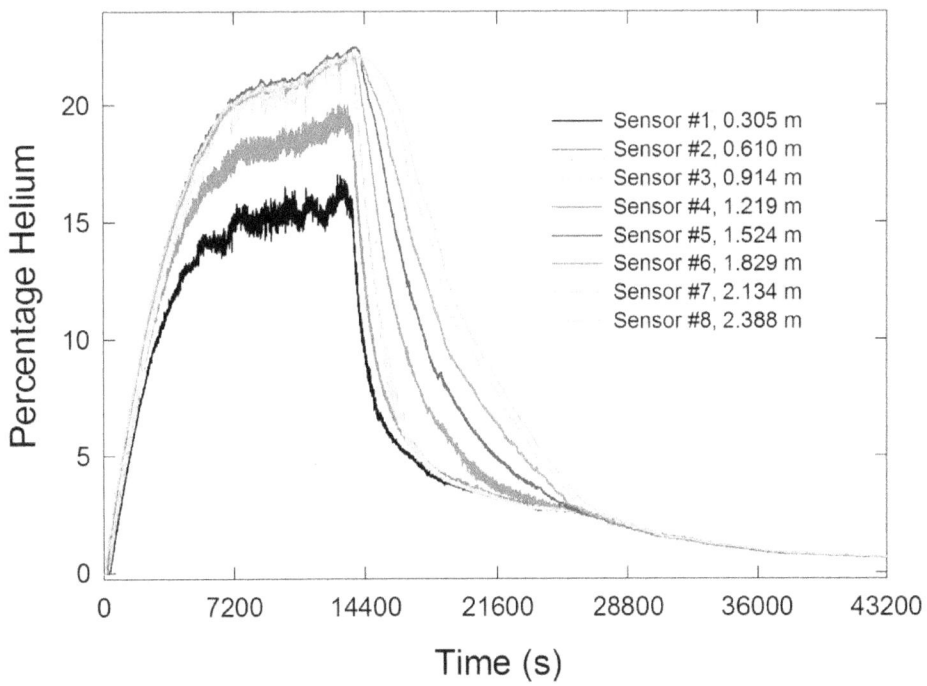

Figure 45. Helium volume percent in the garage is plotted as a function of time at the eight sensor heights indicated along the vertical array for the 8/23/10 experiment.

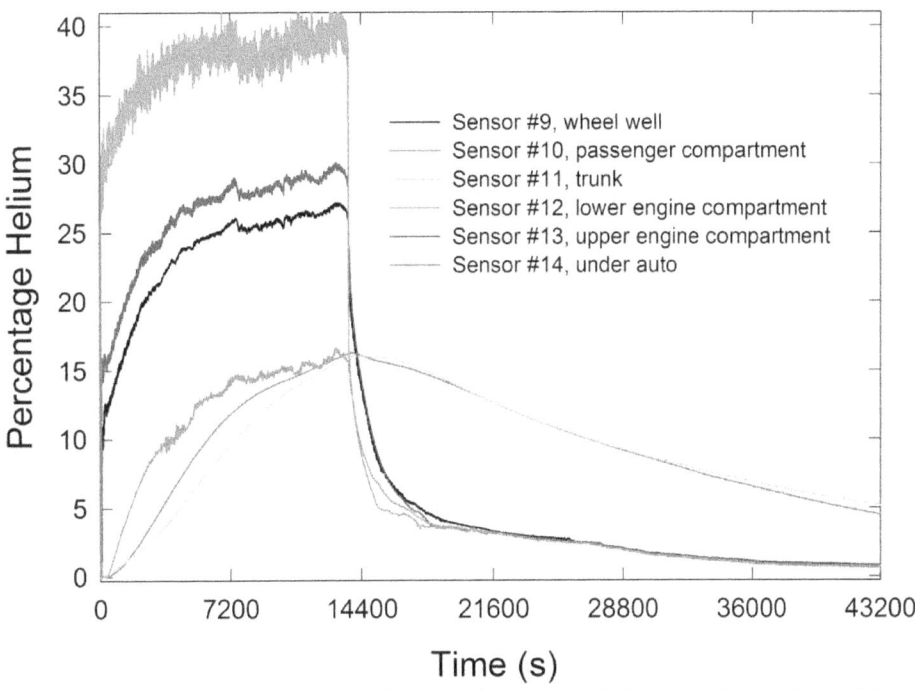

Figure 46 Helium volume percent is plotted as a function of time at the indicated locations in the automobile centered over the helium release location for the 8/23/10 experiment.

Table 6 between the parameters used to characterize these two phases confirms the agreement. This suggests that helium losses from the garage occurred at similar rates for the two experiments. Fits of the

53

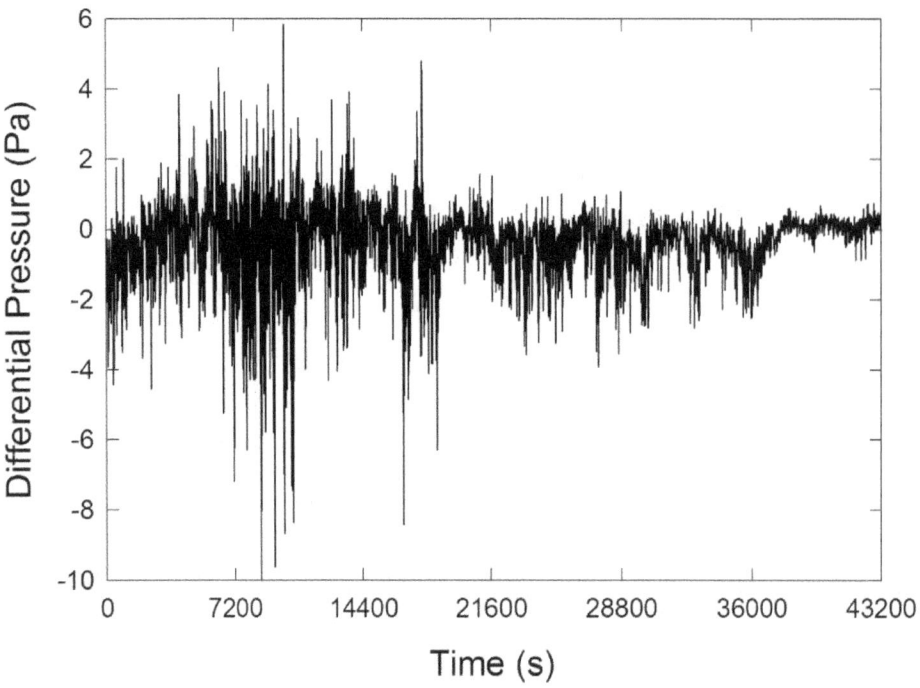

Figure 47. The time variation of the differential pressure between the garage interior and outside ambience is plotted over the time period from 0 s to 43 200 s for the 8/23/10 experiment.

helium concentrations at long times when helium volume percents had fallen to the order of 1 % gave decay constants and ACH_{gar} values that were similar, but the values for 8/23/10 were slightly smaller, indicating slightly reduced gas exchange between the garage and surroundings.

Figure 47 shows a plot of the time variation of the differential pressure between the garage and the outside. There appears to have been a particularly windy period just after 7200 s. There is a leveling off of the helium volume percent curves in Figure 45 and a drop in concentration in Figure 46 that are correlated with the wind increase. The pressure fluctuations and magnitudes on 8/23/10 overall were smaller than observed during the 8/6/10 experiment (see Figure 38). As discussed earlier, these observations suggest that winds of the magnitude present during these experiments have measurable, but generally small, effects on the loss rate of helium from the garage during the release and early post-release periods. The reduced decay rate on 8/23/10 at longer times was likely associated with the lower wind speeds.

The appearances of the concentration profiles for the measurement locations inside the vehicle shown in Figure 46 and Figure 41 were also very similar for the repeated experiments. Detailed comparisons showed that the quasi-steady-state helium volume percent under the vehicle (Sensor #14) reached a slightly higher value on 8/23/10, while those under the hood and in the wheel well were slightly lower. The concentration profiles measured inside the passenger compartment and trunk were nearly identical. Recall that the passenger compartment sensor may have been on the floor instead of at the ceiling for the 8/6/10 experiments. If this were the case, the sensor location had little effect on the measurements.

The decay behaviors of the volume percents following the end of the release were very similar for all measurement locations inside the garage and vehicle. As discussed above, this suggests that helium losses from the garage were comparable in both experiments.

The helium volume percent measured in the family room as a function of time is shown in Figure 48. As in earlier experiments, the concentration began to increase shortly after the helium flow into the garage was started and continued until shortly after the flow stopped. As compared to the earlier experiments (see Figure 17, Figure 23, Figure 29, Figure 35, and Figure 40), the buildup of helium

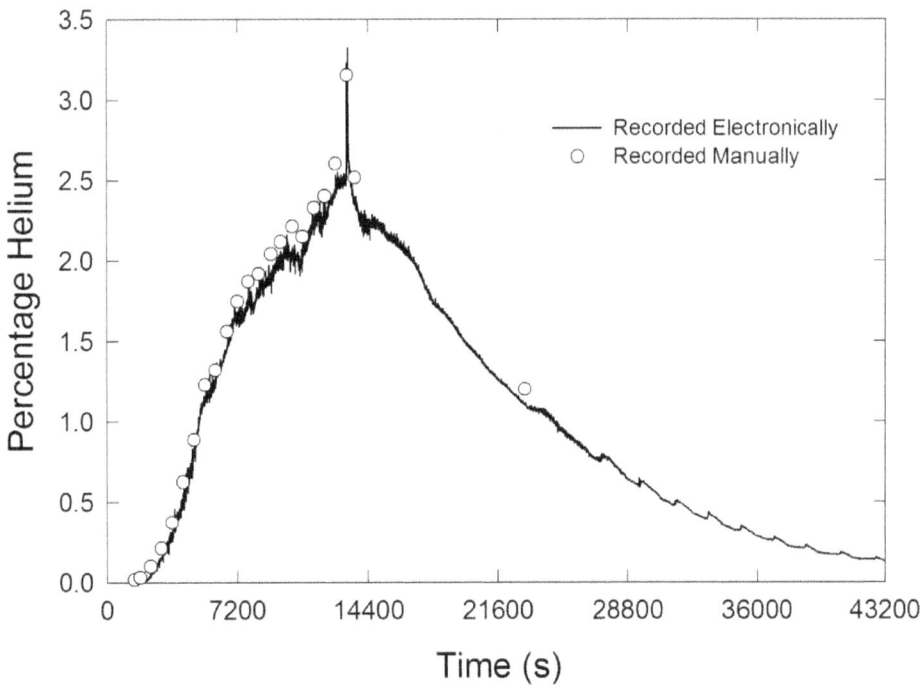

Figure 48. The helium volume percent measured in the family room of the house manually (symbols) and electronically (line) is shown as a function of time for the 8/23/10 experiment.

Table 8. Summary of Forced Ventilation Garage Experimental Flow Parameters

Date	Average He Flow Rate (m³/s)	Flow Period (s)	Total Flow Volume (m³)	Ventilation Fan On (s)	Fan Volume Flow Rate (m³/s)	Vehicle
8/9/10	4.44×10^{-3}	1378	6.1	80	0.0901	Stratus
8/12/10	4.21×10^{-3}	3605	15.2	80	0.0934	Stratus
8/13/10	4.24×10^{-3}	3600	15.3	100	0.0922	Stratus
8/13/10a	4.28×10^{-3}	3600	15.4	101	0.1071	Stratus
8/19/10	4.43×10^{-3}	4130	18.3	100	0.0910	Passat
8/19/10a	4.27×10^{-3}	3600	15.4	138	0.1066	Passat

concentration was considerably reduced, except for the 9/11/08 experiment shown in Figure 17. As discussed earlier, a window was opened during this test, which reduced the interior helium concentration. During the later experiment doors were opened briefly as people entered and left the house, but these were not excessive compared to earlier experiments. The major differences appear to have been that the central air conditioning and fan were not functioning during the 8/23/10 experiment, and the temperature inside the house was somewhat higher. It is interesting that the helium volume percent showed a strong spike when the air conditioning fan was initially turned on when the repairs were completed, increasing by nearly a full percentage point before dropping back down. This is an indication that mixing inside the house was not uniform. This observation does not explain the lower interior concentration measured in the family room. It is considered likely that this was due to a lower exchange rate between the house and garage due to the house fan not operating or the garage-house temperature difference.

3.2 Experiments with Forced Ventilation of the Garage

Six experiments were run during August, 2010 in which the garage was vented with a fan in order to limit the buildup of helium concentration during a release. All experiments were run with a conventional automobile parked above the helium release location at the center of the floor. Table 8 lists the dates of

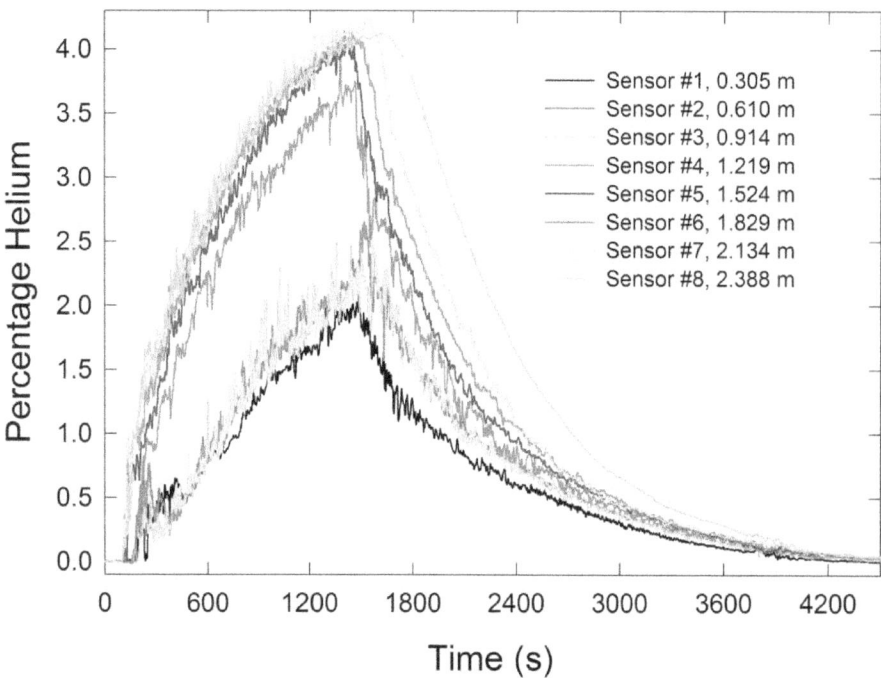

Figure 49. Helium volume percent is plotted as a function of time for the eight sensor heights indicated for the 8/9/10 experiment with forced ventilation.

the experiments, average helium volume flow rates, helium flow periods, total helium flow volumes, the time following the start of the helium flow when the fan was started (roughly corresponds to when the helium volume fraction at Sensor #8 reached 1 %), the fan volume exhaust rates, and the model of the vehicle parked in the garage.

All of these experiments were run during a period when the central air conditioning for the test house was not functioning. Since outside temperatures were generally high during the testing period, interior temperatures inside the house rose significantly. For the last four experiments, a window-unit air conditioner was used to lower temperatures in the family room where electronic equipment was housed, but temperatures elsewhere in the house remained 1 °C to 2 °C higher.

The initial test on 8/9/10 involved a helium release that lasted 1378 s. The exhaust fan was turned on 80 s after the helium flow began. Figure 49 shows plots of helium volume percent versus time for the eight sensors located along the vertical array. Comparison of the concentration profiles with those shown in Figure 36 and Figure 45 for similar helium releases with the Stratus present, but without an exhaust flow, reveals significant modifications of the helium concentration field in the garage due to forced ventilation. Upper-layer helium concentrations for the earlier measurements had increased to around 14 % after 1380 s compared to the roughly 4 % value evident in Figure 49. Helium concentrations were still rapidly increasing without forced ventilation, while they were leveling off with the exhaust fan operating.

The relative vertical concentration profiles were also substantially modified. For the natural ventilation cases, there was a well mixed upper layer above a lower layer where the helium concentration decreased rapidly with height. The transition between the layers was close to Sensor #3 located 91.4 cm above the floor. With forced ventilation, the helium concentration in the lower layer was much more uniform, and the transition height between the layers was slightly higher since the concentration recorded by Sensor #4 at 1.219 m was just barely lower than recorded by the four higher sensors.

When the flow was halted, the helium distribution with forced ventilation responded in the same general way as the tests without a fan operating. Helium concentrations in the lower layer began to fall

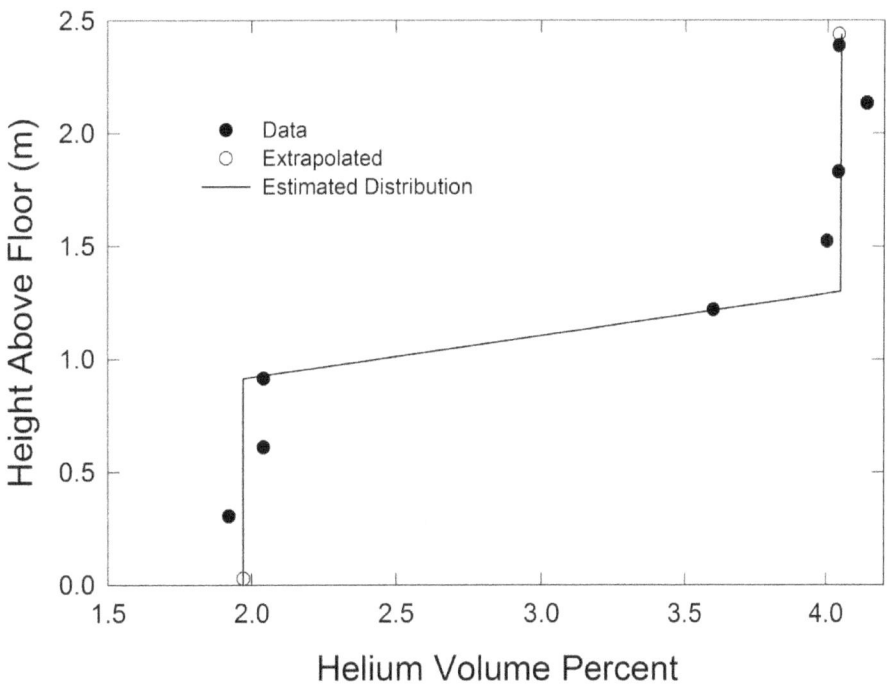

Figure 50. Experimental helium volume percents (solid symbols) measured along the vertical array at the end of the 8/9/10 release are shown along with extrapolated values (open symbols) at the floor and ceiling. The solid lines are approximations of the data used to estimate the average helium concentration along the vertical direction.

immediately, while those in the upper layer first leveled off before starting to decay. Vertical concentration gradients developed, and a substantial period was required for the decaying curves to collapse to a single curve, with concentrations at higher sensors requiring longer periods. A major difference between the behaviors was the substantially faster concentration falloffs with forced ventilation. Roughly 1500 s were required for helium concentrations to approach zero with the exhaust fan operating compared to roughly 60 000 s without forced ventilation.

A number of parameters have been used to characterize the forced-ventilation experiments. These include the upper-layer helium volume fraction at the end of the release period, the average helium concentration along the vertical array at the end of the release period, the period following the end of release required for the helium concentration to fall to 0.2 %, and the helium concentration falloff rate near the end of the experiment when the helium volume fractions dropped from ≈ 0.5 % to ≈ 0.1 %.

The upper-layer concentration was estimated by inspection. The average concentration was determined by an approximate integration performed as follows. Upper and lower concentrations (assumed constant) were estimated visually from the data. The heights were then identified where the constant upper-layer concentration began to fall off and where the constant lower layer concentration began to increase. A linear concentration increase rate was assumed between the two heights. The resulting piecewise linear curve was then easily integrated to give the average concentration. Figure 50 shows an example for the 8/9/10 data. The falloff rate was estimated by least squares fitting an exponential to the experimental data for Sensor #4 over a range where the helium volume percent dropped from 0.5 % to 0.1 %, as shown in Figure 51. The values of these parameters determined for the 8/9/10 test are listed in Table 9.

It was possible to obtain an alternative estimate for the helium volume percent falloff rate by assuming that only air entered the garage to replace gas withdrawn from the interior by the fan, any air that entered was instantly mixed with the interior gas, temperatures were constant, and the exhaust fan

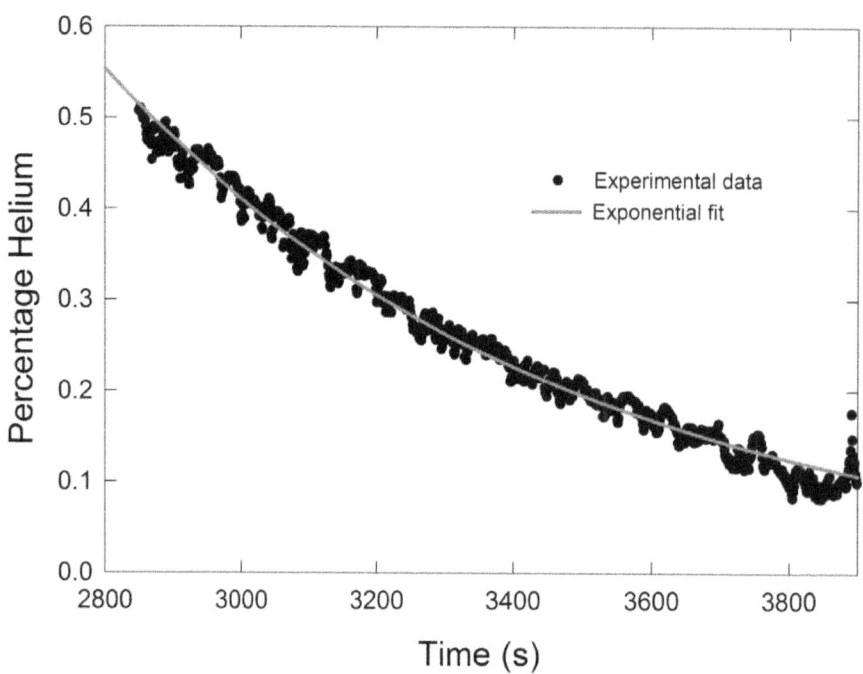

Figure 51. Helium volume percent recorded by Sensor #4 is plotted as a function of time for the 8/9/10 experiment. The red line is the result of fitting the data to an exponential decay using a nonlinear least squares curve fit.

Table 9. Helium Concentration Parameters for Forced-Ventilation Garage Experiments

Date	Upper Layer He Vol. % at End of Release	Average He Vol. % in Garage at End of Release	Time Required for He Vol. % to Drop to 0.2 % at End of Release (s)	Helium Falloff Rate for Sensor #4 at End of Decay (s^{-1})
8/9/10	4.0	3.1	2060	1.49×10^{-3}
8/12/10	4.2	3.3	--	--
8/13/10	4.3	3.4	2030	1.37×10^{-3}
8/13/10a	3.7	2.7	1625	1.81×10^{-3}
8/19/10	4.3	3.7	2350	1.29×10^{-3}
8/19/10a	3.8	3.0	1560	1.86×10^{-3}

efficiency was constant throughout the experiment. For these conditions, the garage helium concentration would have fallen by a factor of two when the amount of exhausted gas was equal to the interior volume, 86.9 m^3, of the garage. Using the exhaust rate provided in Table 8, this would have required 964 s. The corresponding falloff rate is obtained by multiplying the inverse of this time by e$^{-\frac{1}{2}}$ = 0.693. The result is 7.2×10^{-4} s^{-1}. Comparison with the measured falloff rate in Table 9 shows that the measured value was 2.1 times larger. The most likely source of the difference between the two results is the vertical concentration gradient that developed during the post-release period. Since the fan was exhausting gas from the upper part of the garage, the exhausted gas had a higher helium concentration than the average, with the result that helium was extracted faster than predicted assuming instantaneous mixing.

The concentration time profiles measured at five locations inside the car during the 8/9/10 experiment are shown in Figure 52. Corresponding measurements for cases without forced ventilation were provided in Figure 41 (8/6/10) and Figure 46 (8/23/10). The initial concentration increases with forced ventilation were very similar to those observed with natural ventilation. The first detection of helium at the sensor located in the undercarriage of the vehicle (Sensor #14) was 4 s after the flow was initiated, while helium was first detected near the ceiling (Sensor #9) 35 s later. The initial helium volume percent jumps for

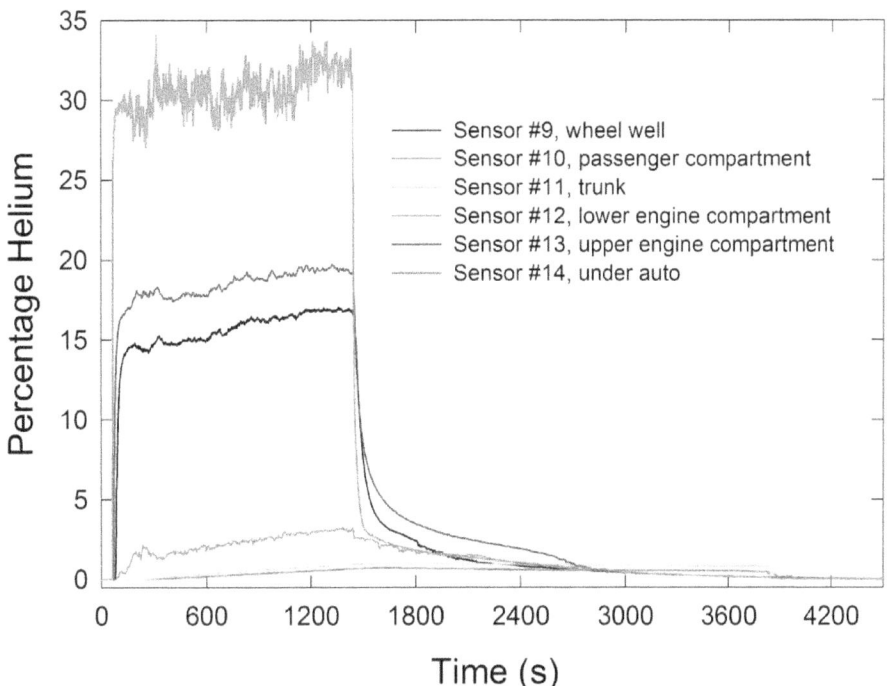

Figure 52. Helium volume percent is plotted as a function of time at the indicated locations in the automobile centered over the helium release location for the 8/9/10 experiment with forced ventilation.

sensors under the automobile and in the engine compartment are indistinguishable in magnitude from the corresponding experiments with natural ventilation. These findings are reasonable since these measurements were recorded prior to the fan being turned on. Note that the helium concentrations trapped under the vehicle correspond to hydrogen concentrations that would be highly flammable if ignited.

The concentrations for Sensors #9, #13 and #14 increased very slowly with time following the abrupt large increases when the helium flow was started. Even though the initial concentration jump for Sensor #12 in the lower engine compartment was much smaller, the slope of the concentration increase with time was similar. The slopes for the concentration increases were noticeably smaller than observed for the natural ventilation cases. This is consistent with the earlier conclusion based on natural-ventilation tests that concentrations at these locations increased as the concentrations exterior to the vehicle increased. The helium concentration at the sensor located near the bottom of the engine compartment (Sensor #12) at the end of the release was nearly a factor of 2 smaller than observed for the cases with natural ventilation at comparable times. When the helium flow was halted, concentrations under the vehicle and in the engine compartment dropped very rapidly to levels equal to the exterior helium volume percent.

As observed for the natural-ventilation experiments, concentrations inside the passenger compartment and in the trunk increased much more slowly during the release period and then fell off much more slowly in the post-release period. Helium volume fraction magnitudes at these locations were slightly reduced at the end of the release period as compared to the natural ventilation cases. This likely reflects the lower helium concentrations trapped under the vehicle and in the engine compartment as well as outside of the vehicle. Once the helium flow was halted, these concentrations fell off very slowly.

Temperatures measured in the family room, inside the garage, and outside the structure during the 8/9/10 experiment are shown in Figure 53. The relatively high temperatures inside the house due to the absence of air conditioning are apparent. Helium volume fractions inside the family room were also

59

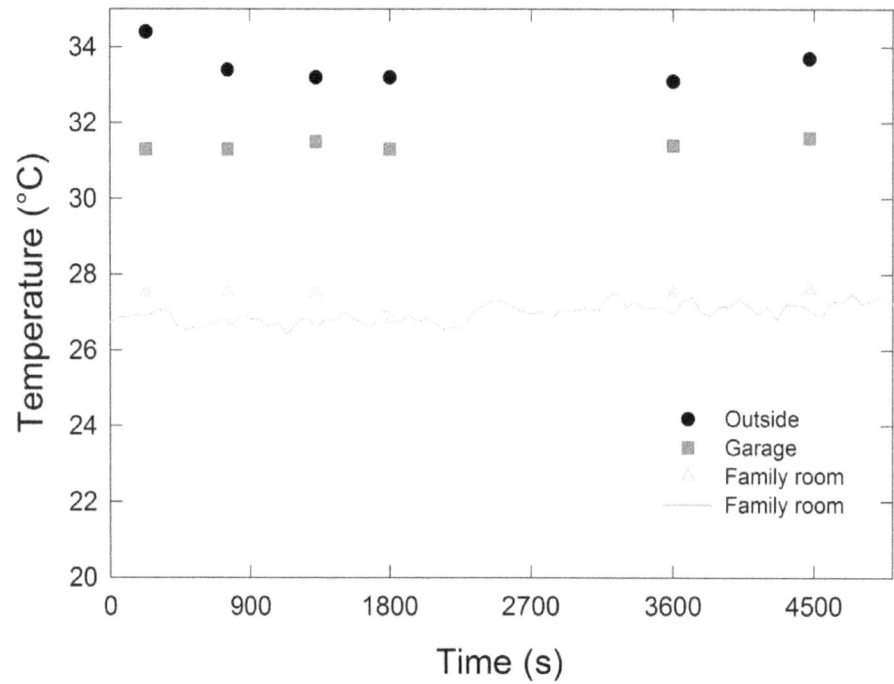

Figure 53. Temperatures measured using thermocouples (symbols) and thermistors (lines) are plotted as a function of time for measurements recorded on 8/9/10 in the family room, garage, and outside.

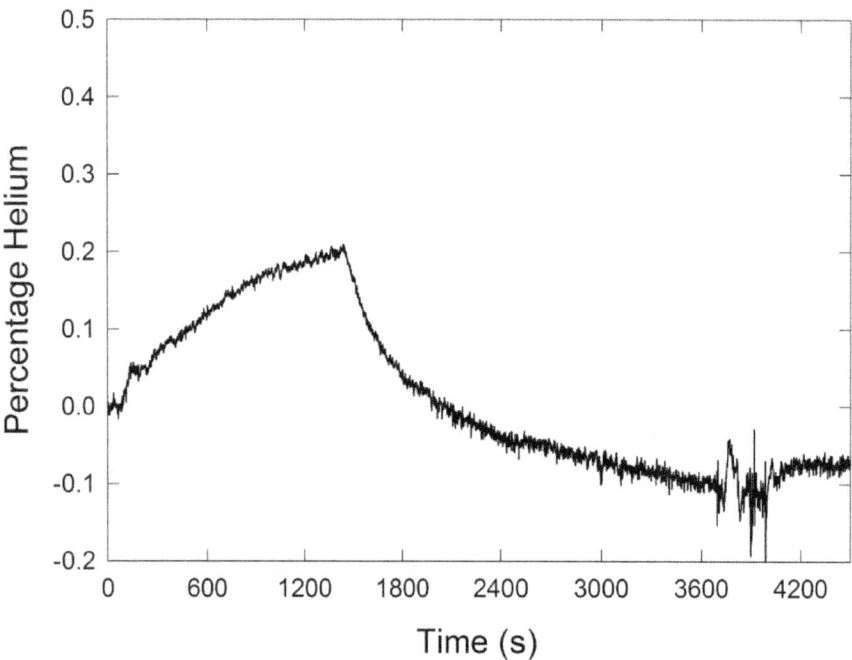

Figure 54. The helium volume percent in the family room of the house measured electronically is shown as a function of time for the 8/9/10 experiment.

recorded and are plotted in Figure 54. A very small increase in helium volume percent was observed during the release period with a maximum concentration of 0.2 % at the end of the release.

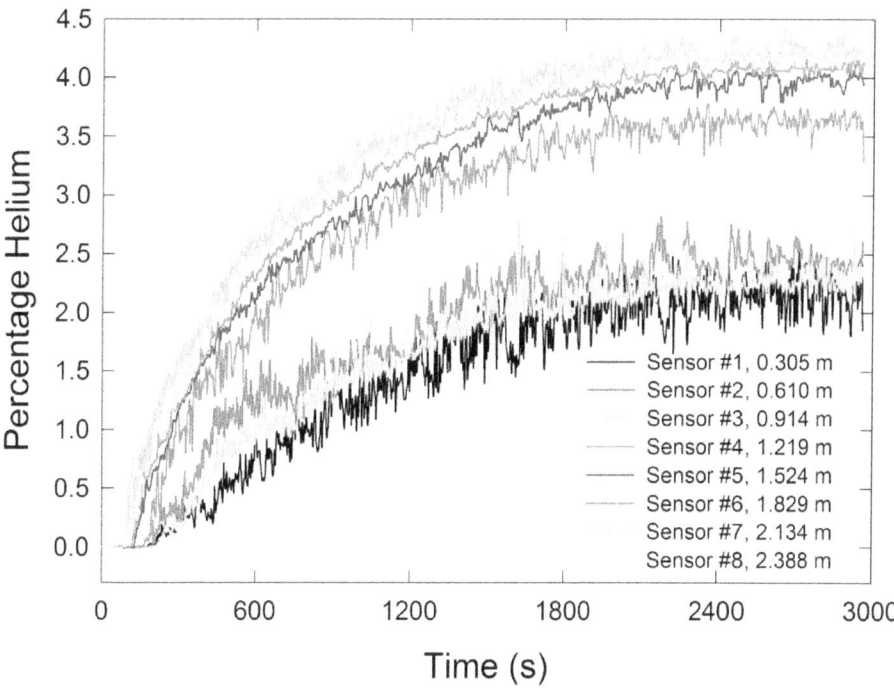

Figure 55. Helium volume percent is plotted as a function of time for the eight sensor heights indicated for the 8/12/10 experiment with forced ventilation.

Figure 55 shows the concentration profiles for locations along the vertical sensor array for the experiment performed on 8/12/10. Once initiated, the helium flow continued until the data acquisition system malfunctioned 2902 s after the flow started. The helium volume flow and garage exhaust rates provided in Table 8 were similar but differed slightly from those used for the 8/9/10 experiment. Comparison of Figure 55 with the concentration profiles for the earlier experiment in Figure 49 shows that the time profiles during the release phase had similar appearances, with two distinct layers present. A detailed comparison of the helium volume percents at the time when the helium flow was halted on 8/9/10 revealed that the helium concentrations were reduced roughly 0.2 % in the lower layer and 0.3 % in the upper layer for the 8/12/10 experiment. The lower values are consistent with both the lower helium volume flow rate and the higher fan exhaust flow rate used during the 8/12/10 experiment (see Table 8).

It is evident in Figure 55 that the helium volume percents continued to rise slowly beyond the time (release period of 1378 s, plotted time of 1438 s) when the helium release was terminated for the 8/9/10 experiment. The helium concentrations did reach asymptotic values after a helium flow period of roughly 2100 s. A detailed comparison showed that the additional concentration increases between 1400 s and 2100 s were 0.4 % at each of the sensor heights. These observations indicate that the upper layer concentration included in Table 8 for the 8/9/10 experiment was likely not the true asymptotic value for this experiment. An estimate for the asymptotic value can be obtained by adding 0.4 % to the listed value to give an upper-layer asymptotic concentration of 4.4 %.

Values of the parameters used to characterize the garage concentration behavior for the 8/12/10 experiment are included in Table 9. Note that no values are included for the post-release phase since measurements were not recorded for this period.

The measured helium volume fractions observed inside and under the vehicle are shown in Figure 56 for the 8/12/10 test. The shapes of the curves at early times as well as the concentration levels are nearly identical to those shown in Figure 52 for the 8/9/10 experiment. Since the release on 8/12/10 lasted for a longer time, it is clear that the helium volume percents at the measurement locations in the engine compartment and immediately under the vehicle were nearly constant following the initial rapid increases

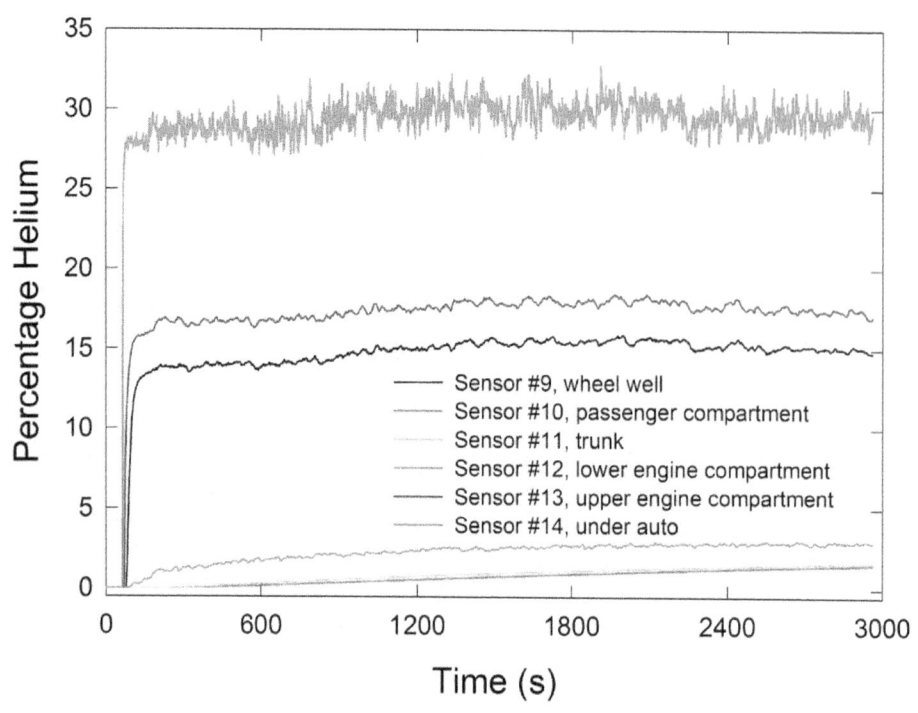

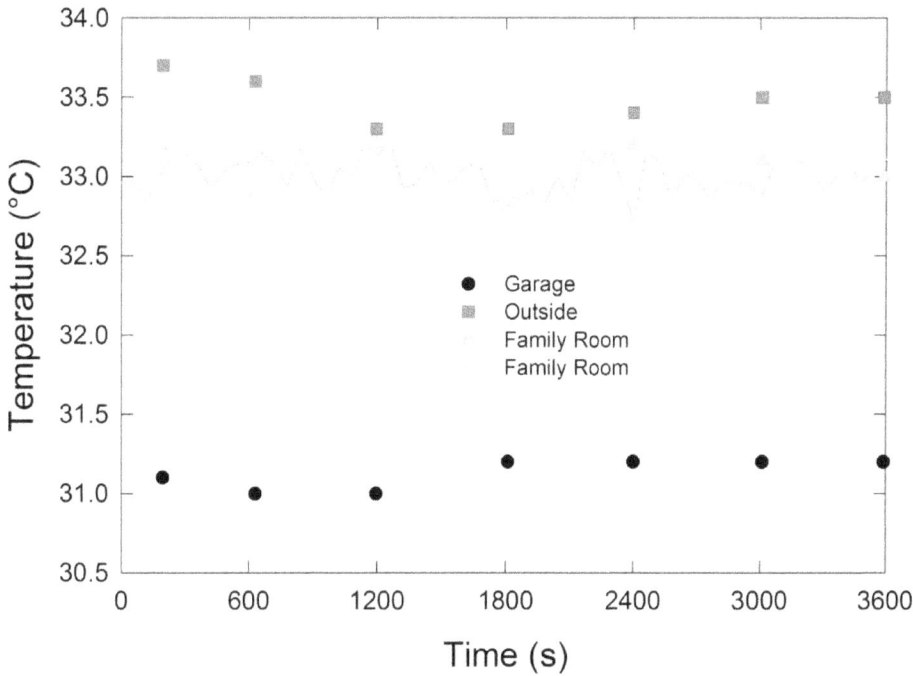

Figure 56. Helium volume percent is plotted as a function of time at the indicated locations in the automobile centered over the helium release location for the 8/12/10 experiment with forced ventilation.

when the helium release was started. The helium concentration in the lower engine compartment at the time when the 8/12/10 flow was stopped was 2.9 %. This value can be compared with the asymptotic value for the lower layer in the surrounding garage of roughly 2.3 %. This observation suggests that a small fraction of the helium in the lower region of the engine compartment came directly from the release. Helium concentrations inside the passenger compartment and trunk rose slowly over the entire release period, reaching levels around 1.6 % at the end of the release. Even after 48 min, these values were well below those trapped in cavities of the vehicle and in the surroundings.

The outside temperature on 8/12/10 was quite hot, and the interior temperature was nearly as high inside the family room as evident in Figure 57. Interestingly, temperatures inside the garage were a couple of degrees cooler. The time profile for the family-room helium concentration is shown in Figure 58. The values appear to have reached a maximum of 0.2 % around 1800 s. This is very close to the value observed on 8/9/10 (see Figure 54). At later times the helium volume percent appeared to drop. It is difficult to know if this drop was real or due to a slow drift of the instrument. The zero shift apparent in the 8/9/10 data suggests that the baseline may have drifted.

Figure 59 shows the helium volume percent time profiles for the vertical sensor array recorded during the first of the two helium releases on 8/13/10. These experiments differed from the earlier two with forced ventilation in that a window air conditioner was used to cool the family room of the attached house, and the weather was considerably cooler. Comparisons with earlier results for 8/9/10 (Figure 49) and 8/12/10 (Figure 55) show that there were marked differences between the earlier concentration profiles and those in Figure 59. Concentrations in the upper layer were roughly 10 % higher for the later experiment, and there were noticeably larger concentration gradients over the upper Sensors #5 to #8. The concentrations at the 1.219 m height (Sensor #4) were much lower relative to the upper-level values than observed for the earlier measurements, indicating a higher interface between the layers, and the concentration fluctuations at this height were increased many fold.

Figure 57. Temperatures measured using thermocouples (symbols) and thermistors (lines) are plotted as a function of time for measurements recorded on 8/12/10 in the family room, garage, and outside.

Figure 58. The helium volume percent measured in the family room of the house electronically is shown as a function of time for the 8/12/10 experiment.

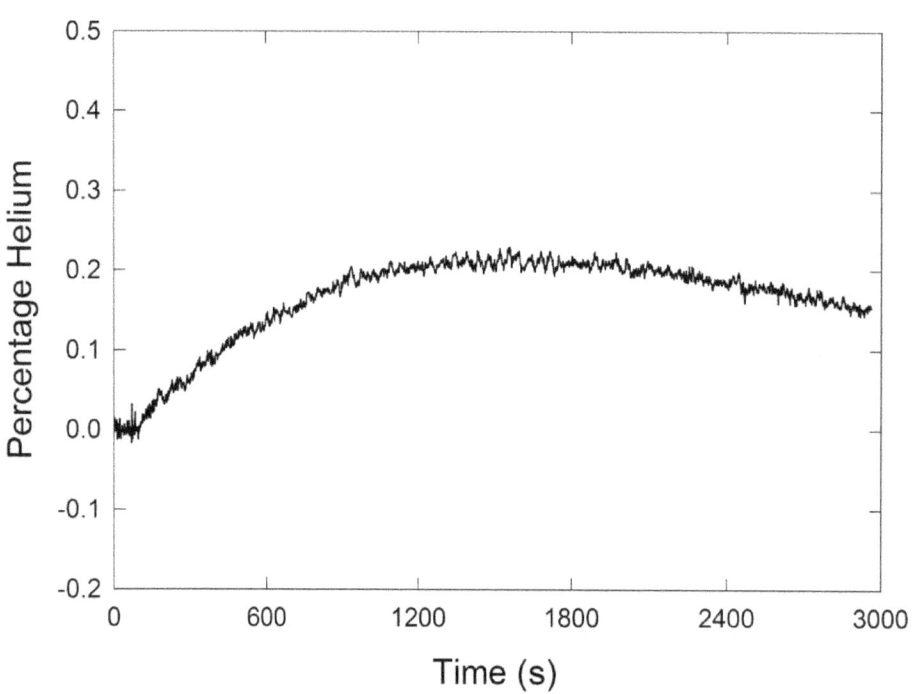

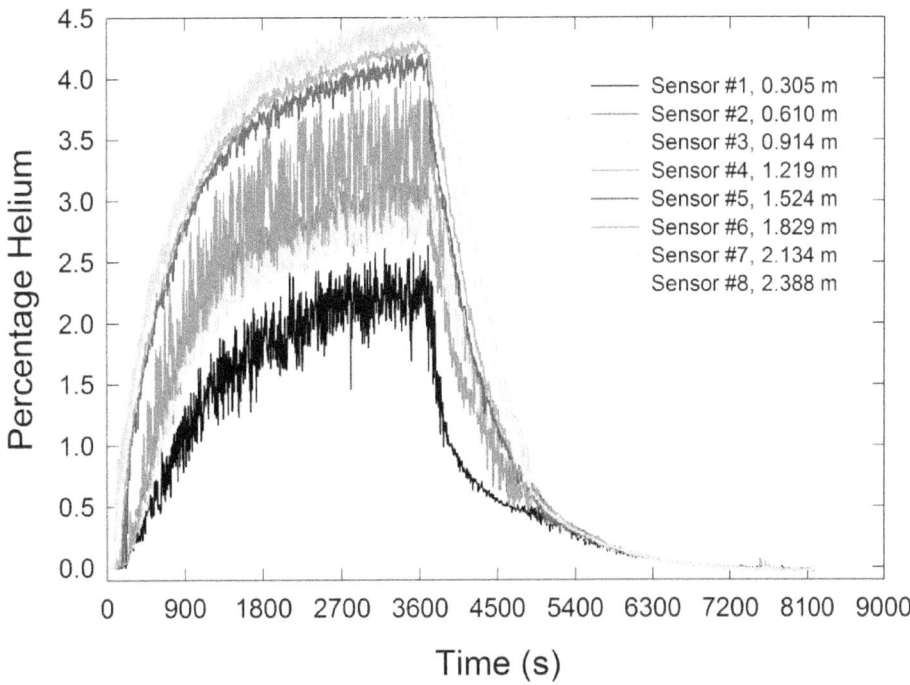

Figure 59. Helium volume percent is plotted as a function of time for the eight sensor heights indicated for the 8/13/10 experiment with forced ventilation.

The concentrations in the lower layer, which were nearly uniform on 8/9/10 and 8/12/10, developed gradients on 8/13/10 with concentrations at the lowest sensor (0.305 cm) lying well below those measured at 0.610 m and 0.914 m. Though close, the helium concentrations measured at 0.61 m (Sensor #2) were slightly higher than recorded at 0.914 m, i.e., there was a negative concentration gradient. This implies that the density at the lower sensor was somewhat less than for the sensor located immediately above. Such a density distribution is said to be "inverted" and creates a condition referred to as "unstable" since higher fluid tends to sink downward. Such a density inversion apparently existed only over a narrow range of heights above the floor. One effect of a density inversion is to increase mixing and concentration fluctuations. It is unclear if the high fluctuation intensities observed at 1.219 m on 8/13/10 were associated with the unstable density distribution.

For the 8/13/10 experiment the helium concentrations rose steeply at all levels during the initial 1500 s of the release and then more slowly. However, unlike the two earlier experiments, the concentrations continued to increase slowly throughout the 1 hr release and did not asymptote as observed earlier. The average concentration over the three lowest sensors seemed to be roughly equal to the asymptoted lower-level uniform values of the earlier experiments.

The substantial changes in vertical concentration distribution between the 8/13/10 data and the earlier experiment on 8/9/10 become clearer when the vertical helium volume percent profile at the end of the release for the later data shown in Figure 60 is compared with Figure 50.

Values used to characterize the post-release phase for the 8/13/10 experiment are included in Table 9. The period required for the helium volume percent to decease to 0.2 % and the concentration decay rate agree well with those for the 8/9/10 data.

There is no apparent correlation between the helium volume or exhaust flow rates listed in Table 8 for the earlier two experiments and the 8/13/10 experiment that provides an explanation for the shift in vertical concentration distribution or higher average concentration. This suggests that these changes must have resulted from a change in one or more external parameters. Figure 61 shows the temperatures measured in the family room, garage, and outside during the 8/13/10 experiment. As already mentioned, the outside temperature was relatively low, and interior and outside temperatures were close to each other.

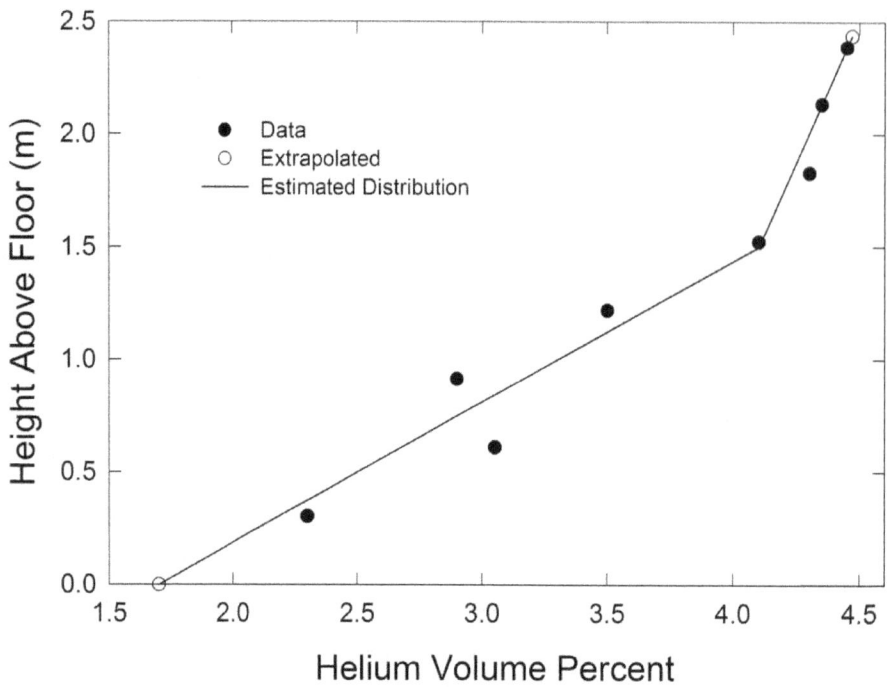

Figure 60. Experimental helium volume percents (solid symbols) measured along the vertical array at the end of the 8/13/10 release are shown along with extrapolated values (open symbols) at the floor and ceiling. The solid lines are approximations of the data used to estimate the average helium concentration along the vertical direction.

Figure 61. Temperatures measured using thermocouples (symbols) and thermistors (lines) are plotted as a function of time for measurements recorded on 8/13/10 in the family room, garage, and outside.

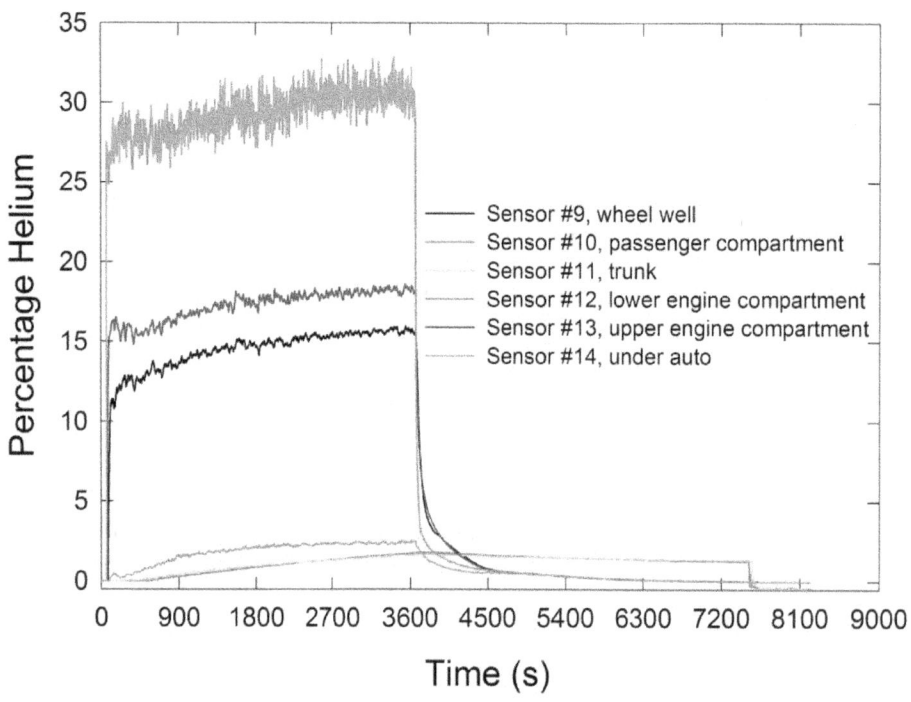

Figure 62. Helium volume percent is plotted as a function of time at the indicated locations in the automobile centered over the helium release location for the 8/13/10 experiment with forced ventilation.

As can be seen in Figure 53 and Figure 57, temperatures were several degrees higher, and the temperature differences between the garage and house were somewhat larger on 8/9/10 and 8/12/10. It seems unlikely that these relatively small differences are sufficient to provide an explanation for such substantial changes in vertical concentration distributions. The helium concentration inside the family room was not recorded on 8/13/10 or 8/19/10.

Unfortunately, wind measurements in the immediate vicinity of the test house were not available for the 2010 experiments. A review of nearby amateur historical weather records for stations located within a few km of NIST did provide some insight. These measurements indicated that 8/9/10 and 8/12/10 were relatively breezy during the experimental periods with measured winds between 2.3 m/s and 10 m/s, while the recorded winds on 8/13/10 were either calm or under 2.3 m/s. On the earlier days, the wind direction was primarily from the southwest, while on 8/13/10 the wind was coming from the northeast. Recall that earlier results without ventilation indicated that wind speed and direction affected helium mixing inside and loss rates from the garage, with higher winds and winds from the south favoring mixing and loss of helium. The observed differences between the 8/13/10 experiment and the two earlier tests are consistent with such wind effects, thus providing a possible explanation for the differences.

Figure 62 shows the helium volume percents recorded by the various sensors placed under and inside the Stratus for the 8/13/10 experiment. The profiles are similar to those observed in the earlier experiments with forced ventilation (see Figure 52 and Figure 56). There are some variations in maximum concentrations achieved and whether or not the values reached asymptotic values.

A second test was run on 8/13/10 that is referred to as 8/13/10a. As can be seen from values included in Table 8, the helium release rates were very close for the two tests, while the fan exhaust rate was increased by about 16 % for the second test. The concentration time profiles along the vertical sensor array are shown in Figure 63 for the 8/13/10a experiment. The relative vertical ordering of the time profiles was similar to those observed earlier in the day (compare with Figure 59), but the profile shapes differed somewhat. For the first experiment the helium concentrations continued to increase slowly until

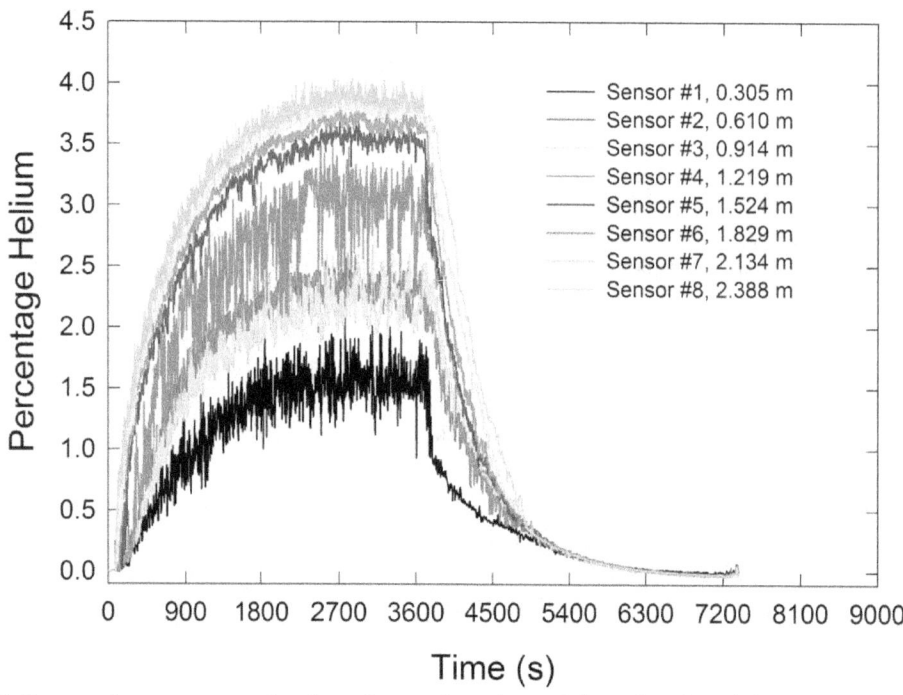

Figure 63. Helium volume percent is plotted as a function of time for the eight sensor heights indicated for the 8/13/10a experiment with forced ventilation.

the end of the helium release following the initial rapid rises, while they reached asymptotic values for the later experiment. More significantly, helium volume percents at the end of the release period for the second experiment were reduced by 0.6 % to 0.7 % at all heights. These reductions in concentration between the 8/13/10 and 8/13/10a experiments are evident by comparing the vertical profiles of the helium volume percents at the end of the release on 8/13/10a shown in Figure 64 with the similar profile in Figure 60 for the 8/13/10 test. The average concentrations along the vertical direction listed in Table 9 for the two experiments confirm this difference, dropping from 3.4 % to 2.7 %. The 25 % decrease in average concentration is larger than the 16 % increase in the fan exhaust rate.

The results indicate that there were higher volume exchange rates between the garage interior and its surroundings for the second experiment on 8/13/10. The increased volume exchange rate was also reflected in the shorter period required for the helium volume percent to drop to 0.2 % during the post-release period for the 8/13/10a experiment, 2030 s versus 1625 s. It should be kept in mind that the average concentrations were different at the ends of the releases. Perhaps more definitive is the observed increase in the helium falloff rate at Sensor #4 as the concentration fell from 0.5 % to 0.1 %, increasing from 1.37×10^{-3} s^{-1} to 1.81×10^{-3} s^{-1} for the 8/13/10 and 8/13/10a experiments, respectively (see Table 9).

Figure 65 is a plot of the temperatures inside the house and garage and outside during the period of the 8/13/10a experiment. The outside temperature had risen slightly from the time of the earlier experiment (see Figure 61), while the temperature inside the house had fallen. The garage temperatures remained roughly constant. Given the small changes, it seems unlikely that temperature variations provide an explanation for differences between the two 8/13/10 experiments. Wind measurements recorded by nearby amateur weather stations indicated that conditions remained relatively calm during both experiments on 8/13/10. These observations lead to a conclusion that differences in vertical helium concentration distributions at the end of release periods for the two 8/13/10 experiments were the result of the change in fan exhaust flow rate.

The helium concentrations measured at the various locations inside the Stratus during the 8/13/10a experiment are plotted in Figure 66 and can be compared with those for the 8/13/10 test in Figure 62. The two sets of time profiles have very similar time dependencies and magnitudes. Careful comparison

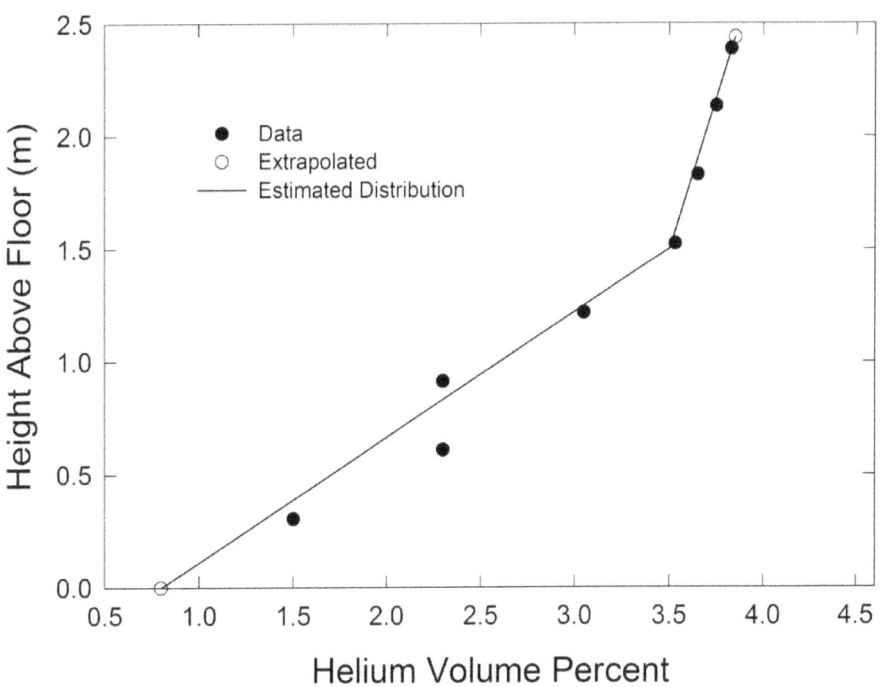

Figure 64. Experimental helium volume percents (solid symbols) measured along the vertical array at the end of the 8/13/10a release are shown along with extrapolated values (open symbols) at the floor and ceiling. The solid lines are approximations of the data used to estimate the average helium concentration along the vertical direction.

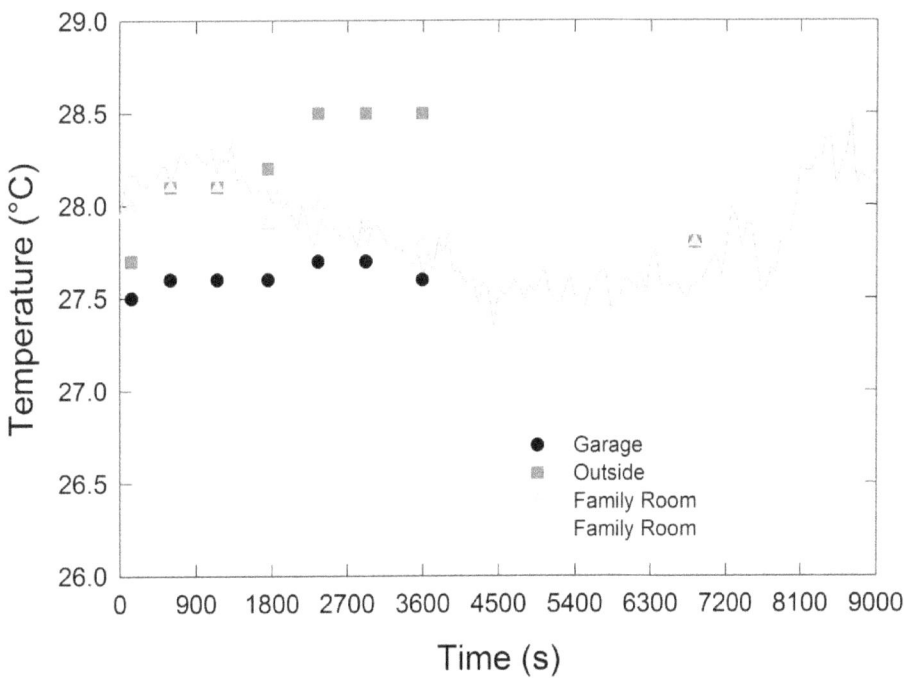

Figure 65. Temperatures measured using thermocouples (symbols) and thermistors (lines) are plotted as a function of time for measurements recorded on 8/13/10a in the family room, garage, and outside.

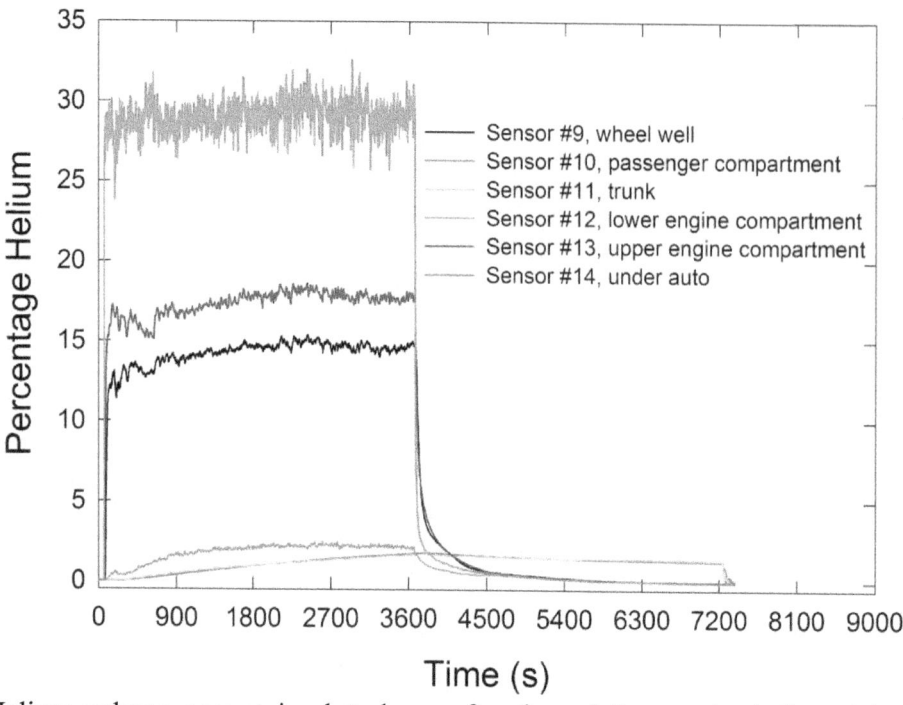

Figure 66. Helium volume percent is plotted as a function of time at the indicated locations in the automobile centered over the helium release location for the 8/13/10a experiment with forced ventilation.

reveals that, similar to the profiles along the vertical sensor array in the garage, the profiles under the auto and in the engine compartment achieved asymptotic values for the later experiment, while the same profiles recorded earlier continued to increase slowly over the entire release phase. These small differences are likely due to the changes in helium volume percents immediately outside the vehicle. As discussed previously, the surrounding helium concentration distribution influences those inside the vehicle due to the strong mixing of released helium with outside gas. Since the concentrations exterior to the vehicle and their rate of increase were small, their influence was relatively weak.

Two additional experiments, 8/19/10 and 8/19/10a, were run during which forced ventilation was used. For these tests the Stratus was replaced with a 2003 Volkswagen Passat. The underside of the Passat has a different configuration than the Stratus, which is open underneath the vehicle. Much of the undercarriage of the Passat is protected by plastic splash shields which extend down to the level of the outer edge of the vehicle. The shields do not extend all of the way to the center of the undercarriage, and there is a narrow recessed channel containing the engine exhaust lines running along the center from the gasoline tank at the rear and connecting into the engine compartment at the front. The different underside configuration of the Passat is likely to have directed more helium directly to the sides of the vehicle. It was not possible to place a helium sensor under the Passat in the same location as for the Stratus. As an alternative, it was centered on top of the left front wheel strut 18.4 cm above the floor of the garage.

The concentration profiles for locations along the vertical sensor array are shown in Figure 67 for the 8/19/10 test. Note that this release lasted 4130 s, 530 s longer than the 3600 s periods used for the two 8/13/10 tests, and the helium volume flow rate was roughly 4 % higher. The concentration profiles had behaviors that were intermediate between those observed for the two forced-ventilation experiments on 8/9/10 and 8/12/10 and the two experiments on 8/13/10. Similar to the first two experiments, the upper layer was well mixed, and the helium concentration for the 121.9 m height was close to the upper layer value, even thought the concentration fluctuations were greater than observed for these experiments. On the other hand, the lower layer developed concentration gradients, but these were somewhat reduced

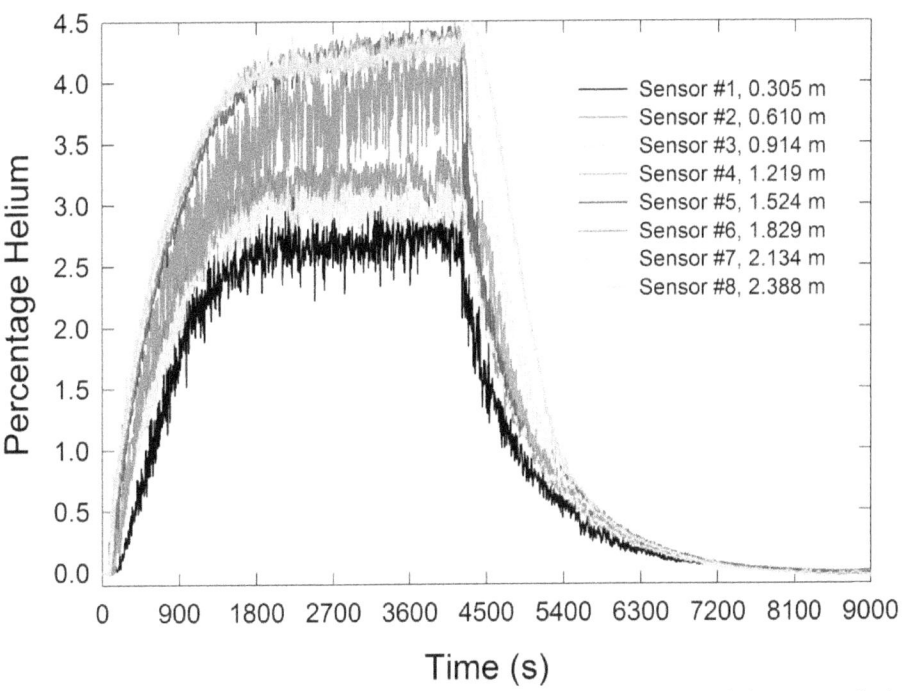

Figure 67. Helium volume percent is plotted as a function of time for the eight sensor heights indicated for the 8/19/10 experiment with forced ventilation.

compared to those observed on 8/13/10. The weak density inversion was still evident between Sensor #2 and Sensor #3. Figure 68 shows the experimental helium volume percent values at the end of the release plotted as a function of height. The lines are approximations for the concentration distribution that were used to estimate the average value of 3.7 % included in Table 9.

Temperatures recorded during the 8/19/10 experiment are shown in Figure 69. The temperature differences between the garage and the house interiors and the outside ambient were not very large. It is unlikely that these small variations were responsible for observed differences between experiments.

The average helium volume percent estimated for the 8/19/10 experiment was the largest of the five experiments discussed thus far that employed an exhaust fan. This was likely due to the higher helium volume flow rate and longer release period used for this experiment. With the exception of the 8/13/10a experiment, the exhaust volume flow rates were similar to those used in the earlier measurements.

The concentration profiles at locations inside the Passat for the 8/19/10 experiment are shown in Figure 70. The profiles appear very different than earlier experiments primarily because Sensor #14, which was located under the vehicle, recorded a maximum helium volume percent around 1.5 % instead of the roughly 30 % measured when the tests were run with the Stratus parked in the garage. This large difference was due to the change in the sensor location mentioned above. The placement of the sensor on the wheel strut apparently moved it out of the layer of helium trapped underneath the car. The helium concentration measured by this sensor overlaps that recorded by Sensor #12, which was located near the bottom-front of the engine compartment at the base of the radiator. Helium volume percents recorded in the upper part of the engine compartment and in the passenger side front wheel well had similar time behaviors and magnitudes to those observed for the Stratus, but the levels at the top of the engine compartment were somewhat higher for the Passat. Helium concentrations inside the passenger compartment and trunk of the Passat grew in a similar way as observed with the Stratus, increasing slowly during the helium release and then decaying much slower than observed inside the garage or under the vehicle. After 4130 s of helium release, the volume percent in the passenger compartment and trunk had reached levels of 2.1 % and 2.7 %, respectively. Comparable values were 1.7 % at both locations for

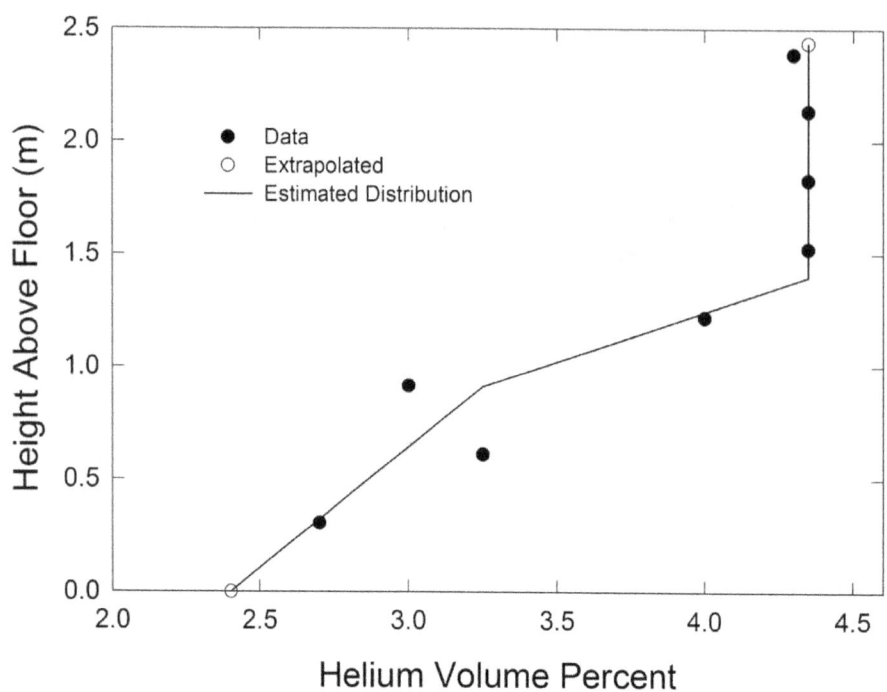

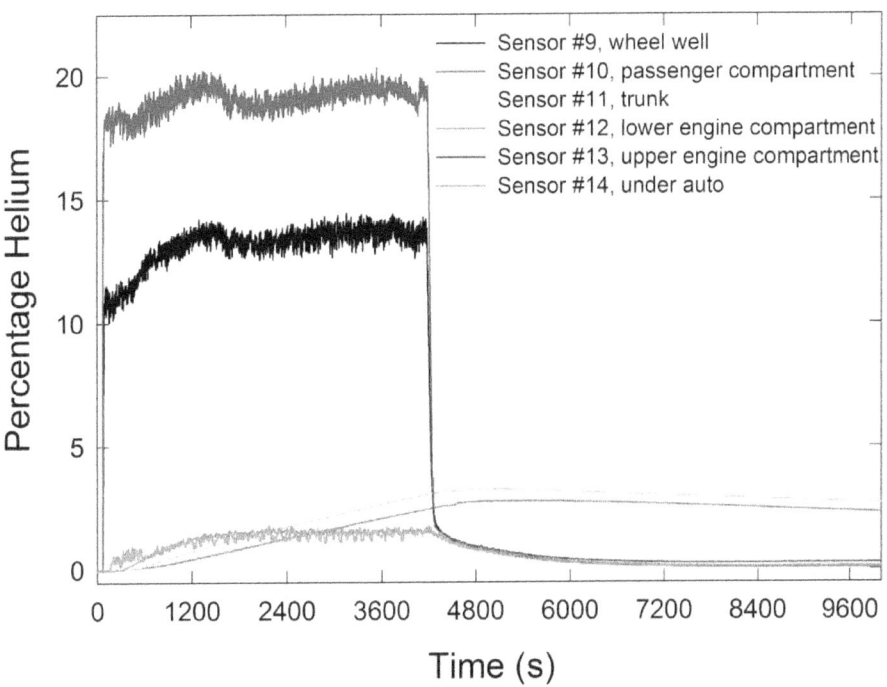

Figure 68. Experimental helium volume percents (solid symbols) measured along the vertical array at the end of the 8/19/10 release are shown along with extrapolated values (open symbols) at the floor and ceiling. The solid lines are approximations of the data used to estimate the average helium concentration along the vertical direction.

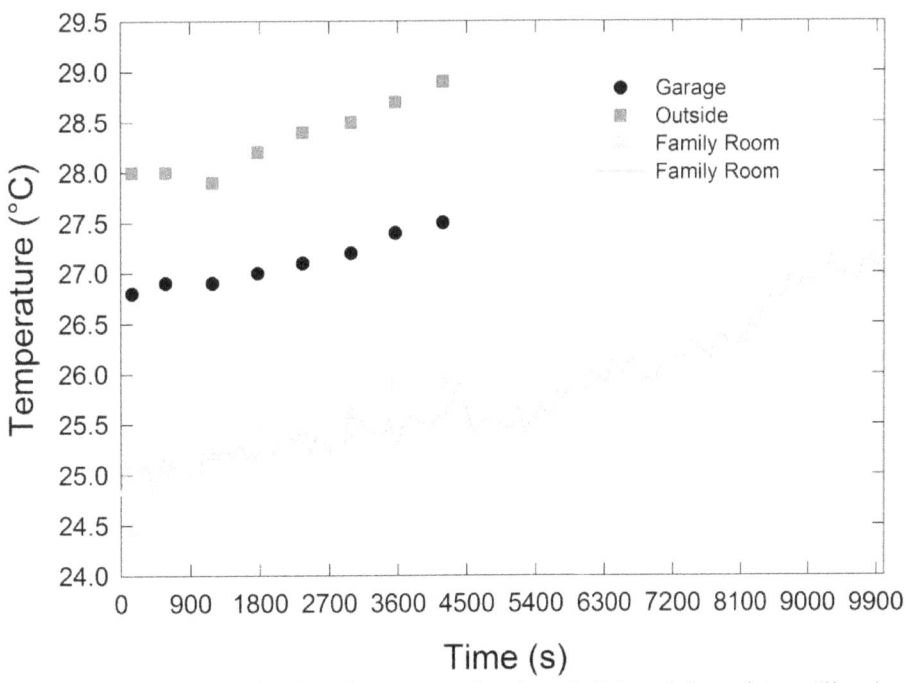

Figure 69. Temperatures measured using thermocouples (symbols) and thermistors (lines) are plotted as a function of time for measurements recorded on 8/19/10 in the family room, garage, and outside.

Figure 70. Helium volume percent is plotted as a function of time at the indicated locations in the Passat centered over the helium release location for the 8/19/10 experiment with forced ventilation.

the two 8/13/10 experiments. Apparently, the transport of helium between the surroundings and the compartments was slightly higher for the Passat.

The helium concentration profiles along the vertical sensor array are shown in Figure 71 for the second test, 9/19/10a. This experiment employed a 17 % higher exhaust flow rate than the earlier test (see Table 8). Comparison with Figure 67 reveals substantial differences between the two vertical profiles. The later data yielded a vertical concentration distribution which appeared to be very similar to those recorded on 8/9/10 and 8/12/10. A fairly well mixed upper layer (a weak concentration gradient is evident) and well mixed lower layer were present in contrast to the nonuniform lower layer observed earlier in the day. Helium volume percents recorded at Sensor #4 (1.219 m) were closer to the upper layer concentration and had fluctuations that were substantially reduced.

The change in vertical concentration profiles can be seen more clearly by comparing the profile at the end of the release for the 8/19/10a data shown in Figure 72 with that for 8/19/10 in Figure 68. The solid lines in Figure 72 are an approximation to the concentration distribution which was used to estimate the average concentration at the end of the helium release of 3.0 % listed in Table 9. The average value is 23 % smaller than the value estimated for the 8/19/10 data. A similar reduction of 26 % was obtained when the exhaust volume flow rate was increased by a comparable value for the two experiments on 8/13/10. The fall off rate for Sensor #4 over the 0.5 % to 0.1 % helium volume percent range for the 8/19/10a test is included in Table 9. This value is in excellent agreement with that measured for the 9/13/10a experiment, which was run using a nearly identical exhaust flow rate.

The first experiment on 8/19/10 was run in the morning. The second was started in early afternoon. As is evident by comparing Figure 73 with Figure 69, the outside temperature rose rapidly over the day. Even though the temperature was rising, there was only a 2.5 °C difference between the garage temperature, which was only slightly warmer than the family room in the house, and the outside. Pressure differences generated as the result of these temperature differences were likely negligible compared to those associated with the fan-induced flow.

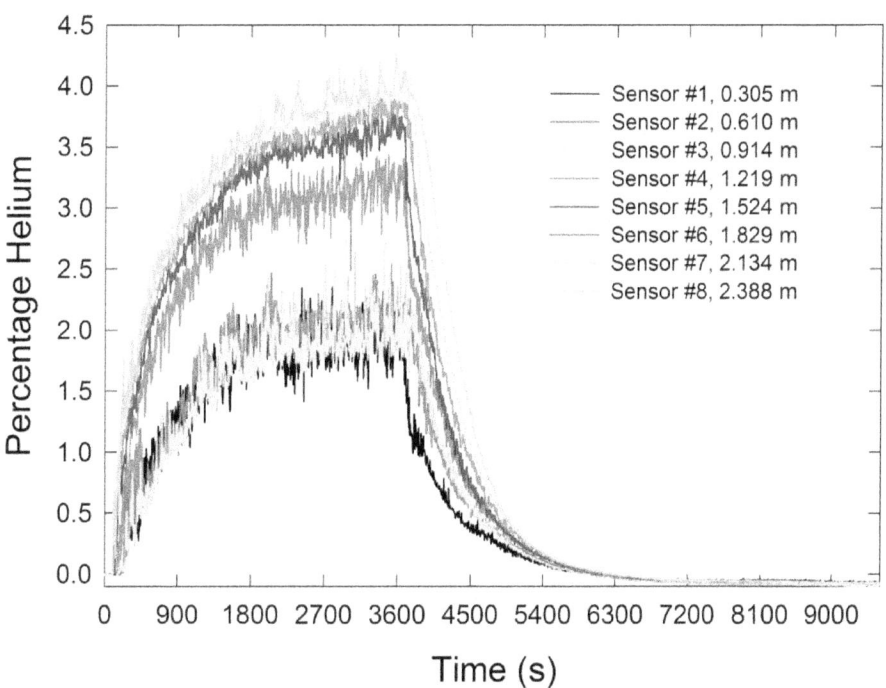

Figure 71. Helium volume percent is plotted as a function of time for the eight sensor heights indicated for the 8/19/10a experiment with forced ventilation.

Figure 72. Experimental helium volume percents (solid symbols) measured along the vertical array at the end of the 8/19/10a release are shown along with extrapolated values (open symbols) at the floor and ceiling. The solid lines are approximations of the data used to estimate the average helium concentration along the vertical direction.

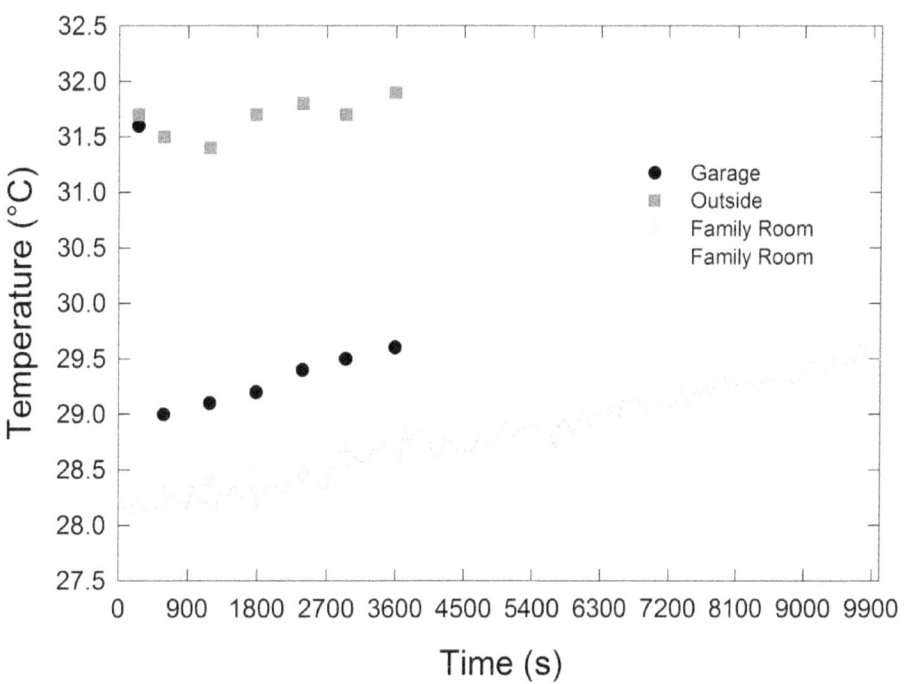

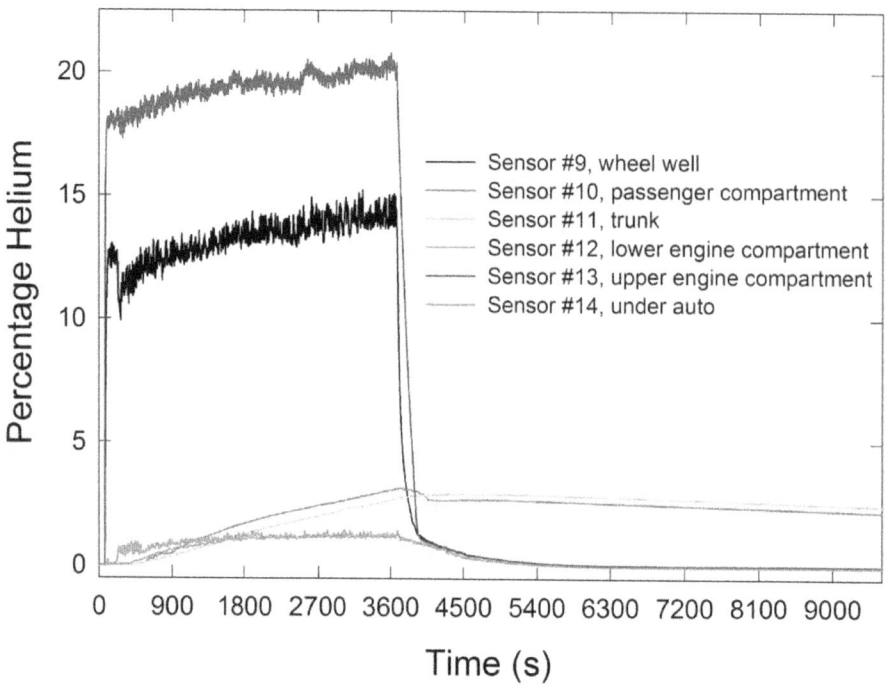

Figure 73. Temperatures measured using thermocouples (symbols) and thermistors (lines) are plotted as a function of time for measurements recorded on 8/19/10a in the family room, garage, and outside.

The helium volume percent time profiles measured at various locations under and within the Passat are plotted in Figure 74 for the 8/19/10a experiment. The profiles are similar to those measured for the same locations in the morning and shown in Figure 70, but there are some subtle and interesting differences. For both experiments, the helium volume percents recorded in the lower part of the engine compartment (Sensor #12) and under the vehicle (Sensor #14) overlap at the end of the release period. However, the approximate values at this time were reduced from 1.6% to 1.2 % for the later test. The measurements inside the passenger compartment and trunk also differed slightly. Measurements in the trunk were the same after 3600 s of helium release for the two releases, reaching values of 2.7 %. However, as can be seen in the figures, the concentrations in the passenger compartment were lower, 2.1 %, for the 8/19/10 run and higher, 3.1 %, for 8/19/10a. Interestingly, for the latter case, the helium concentration in the passenger compartment dropped relatively quickly at the end of the release period and fell below the value in the trunk so that the long period of slow decay looked very similar to that for the 8/19/10 test. It is difficult to explain the reasons for these observations, but they were likely associated with the different vertical concentration distribution in the surrounding garage.

The change in vertical concentration distribution between the two experiments on 8/19/10 are similar to that identified between the 8/9/10 and 8/12/10 experiments and the two tests performed on 8/13/10, with the 8/19/10a results being most similar to the results on 8/9/10 and 8/12/10. Wind effects have been tentatively identified as having been at least partially responsible for the variations among the earlier measurements. Local weather conditions recorded by several nearby amateur weather stations for 8/19/10 were reviewed. There were inconsistencies between stations, but these historical records generally indicated increasing winds between the two experiments on 8/19/10, with the wind shifting from the north to the west. These observations are consistent with a conclusion that higher winds from the south and west resulted in more uniform concentration distributions in the garage during experiments with active ventilation.

Figure 74. Helium volume percent is plotted as a function of time at the indicated locations in the Passat centered over the helium release location for the 8/19/10a experiment with forced ventilation.

4 Discussion

4.1 Naturally Ventilated Garage Experiments

Six experiments with natural ventilation of the garage performed over a two year period have been described. Both momentum- and buoyancy-dominated helium flows were released into the garage. Two experiments were done with a vehicle parked over the buoyancy-dominated helium source. Helium volume percents were recorded as a function of time and height at a location centered in the left-rear quadrant of the garage. Figure 9, Figure 18, Figure 24, Figure 30, Figure 36, and Figure 45 show plots of helium volume percent for the six tests spanning times covering the helium release and post-release periods. Comparisons showed that the general appearances of the concentration time behaviors were similar for all of these tests.

At the start of a release there were relatively rapid increases in measured concentrations, but after roughly two hours the rates fell dramatically, and the concentration profiles developed quasi-steady-state behaviors that persisted until the helium flow was halted. The development of common curves for the highest sensors showed that well-mixed upper layers developed during each of the experiments. The quasi-steady-state helium volume percents for each experiment are listed in Table 5. The values are similar for each test, yielding an average helium volume percent of 22.6 % with a standard deviation of 0.4 %. The standard deviation is only slightly larger than the estimated uncertainty of 0.3 % for an individual test. Comparisons of quasi-steady-state upper-layer helium volume percent profiles with time records of wind speed and direction and differential pressure between the garage interior and outside indicated a weak dependency on these parameters that was sufficient to account for variations of this magnitude. The 2008 experiments had slightly higher quasi-steady-state concentrations, but the differences were small and of the same order of magnitude as the variations due to changing wind conditions. The data were checked to see whether there was an identifiable dependence on the helium volume flow rate or total volume of helium released. No such dependence was identified.

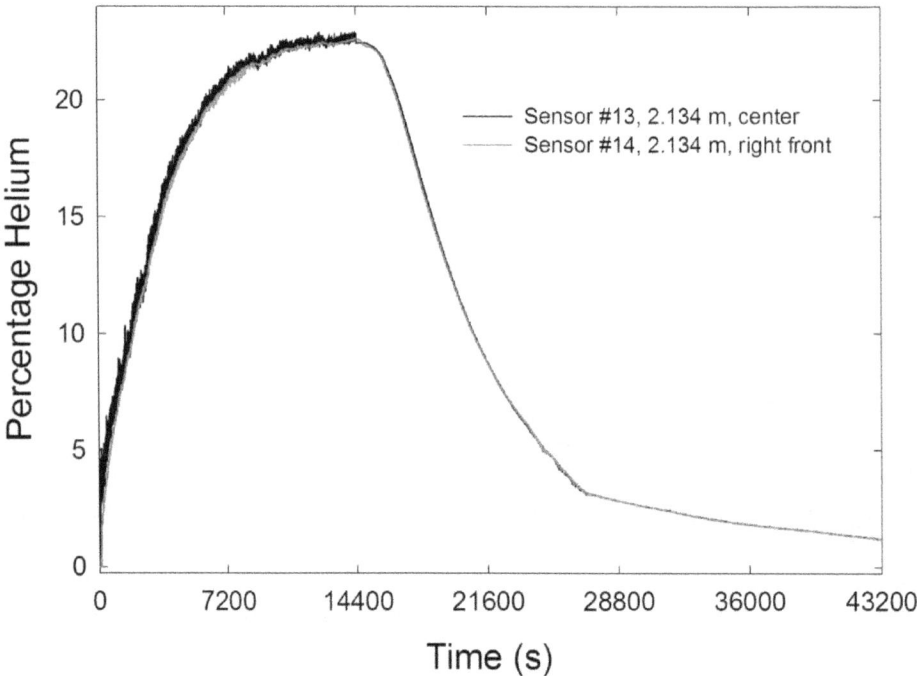

Figure 75. Helium volume percent is plotted as a function of time for sensors located 2.134 m above the floor immediately above the helium release location and in the front-right quadrant of the garage for the 7/29/10 experiment.

In our earlier reduced-scale study, it was shown that horizontal gradients were quickly smoothed out by flows that developed when vertical density profiles varied across the garage. [1] As a result, the measurements recorded along the vertical array should be representative of the vertical helium volume percent profiles elsewhere in the garage. This assumption was used to estimate quasi-steady-state average helium volume percents for the entire garage.

The hypothesis that vertical concentration profiles did not vary with lateral location was checked using measurements made along the flow release vertical centerline and in the front-right quadrant of the garage during the 7/29/10 and 8/2/10 experiments (see Figure 26 and Figure 31). In these figures it is evident that the concentrations measured at 1.219 m and higher above the floor along the plume centerline and at the right-front height of 2.134 m had similar values. The fluctuations and slightly higher helium concentrations present along the plume axis introduce some ambiguity during the releases, but, as evident in the figures, these fluctuations dissipated, and the local concentrations rapidly recovered immediately after the helium flow was halted. Inspection showed that at these times all of the helium volume fractions at the upper-layer locations above the release point had nearly identical values. Using the results for locations 2.134 m and higher, characteristic concentrations were obtained in the same way as for measurements along the vertical array. The results were 22.5 % ± 0.1 % for the 7/29/10 experiment and 22.9 % ± 0.1 % for the 8/2/10 experiment. Note that the small variations observed indicate that the upper-layer helium volume fractions above the release point and in the front-right quadrant of the garage were identical. Comparison with the results in Table 5 shows that the estimated helium volume percents for the center and right-front upper layer were 0.2 % higher for both experiments. This difference is well within the variations observed and indicates that the assumption that horizontal concentration gradients were negligible is valid.

Additional confirmation of a horizontally uniform upper layer is obtained by directly comparing the 7/29/10 helium concentrations recorded by the sensors located 2.134 m above the helium release location and at the same height in the right-front quadrant of the garage in Figure 75. The slightly higher concentration and fluctuations above the release location as compared to the right-front position were due to the presence of the plume during the helium release. As soon as the helium flow stopped, the curves for the two locations became completely overlapped and remained so as the helium concentration fell off.

Average quasi-steady-state helium volume percent values for the garage are listed in Table 5. The average for the six tests was 19.6 % with a standard deviation of 0.6 %. The variations between tests are relatively small, but larger than the variation in the uniform upper-layer values. While the two highest quasi-steady-state upper-layer helium volume percents were recorded for the 2008 experiments with helium release from a tube, the corresponding average values for the garage were the lowest. Comparison of the vertical concentration profiles indicates the differences were at least partly due to variations in concentration gradients in the lower part of the garage. Due to the spacing of the sensors, it was difficult to identify the exact location of the interface between the uniform upper layer and the lower region where the concentration fell off. Comparisons of the vertical profiles for the 2010 experiments showed that the interface did move lower in the garage for the experiments with a vehicle present as compared to those without. The average concentration in the garage also changed as the interface height varied.

The measurements indicate that uniform upper layers formed during the helium releases for all six experiments. Based on the Froude numbers, the helium flows for the experiments without a vehicle in 2008 were momentum dominated when they reached the ceiling, while those in 2010 were buoyancy dominated. Both types of flow formed uniform upper layers with interfaces located near 0.9 m. Cariteau et al. [13], Merilo et al. [16] and Ekoto et al. [18] reported that momentum-dominated upward flows favor the formation of well mixed upper layers, while buoyancy-dominated flows favor the formation of stratified upper layers. These findings differ from the current work where both types of helium sources generated well mixed upper layers.

The concentration measurements along the centerline of the release for the experiments without a vehicle provide insight into the reason for the different conclusion. These measurements (see Figure 26, Figure 31, and Figure 75) reveal that the plume rapidly entrained surrounding gas and was highly mixed by the time it reached the sensor located 0.67 m above the floor and that helium concentrations inside and outside the plume were very similar. In contrast, Cariteau et al. showed that sensors placed near the ceiling inside their buoyancy-dominated flows recorded significantly higher helium concentrations than observed for locations at the same height outside the plume. [13] The different plume mixing behaviors provide the likely explanation for why the upper layers displayed different stratification behaviors.

Cariteau et al. considered helium flows having $Fr = 0.44$ and 1.29. [13] These values are considerably higher than the $Fr = 0.075$ used for the releases from the 0.305 m × 0.305 m flow conditioner in the current experiments. Merilo et al. reported that the Froude number for their buoyancy-dominated flow was 62. [16] This value suggests that the flow was momentum-dominated at the release point. However, Merilo et al. argued that the flow became primarily buoyancy dominated by the time the flow reached the ceiling. The corresponding value for their momentum-dominated case was $Fr = 664$. Apparently, the development of upper-layer stratification within a garage depends strongly on Fr, with maximum stratification occurring for Fr around one and more complete mixing occurring for Fr much greater than or much less than one.

The results in Table 5 indicate that slightly more than 2/3 of the volume of helium released over a roughly 4 h period was lost from the garage during the releases. Since quasi-steady-state helium volume percents, $HeVol\%_{ss}$, were achieved, it is possible to estimate the effective air changes per hour rate, $ACH_{eff,ss}$, for the garage at the end of the helium release period. The helium volume percent for the quasi-steady state can be written as

$$HeVol\%_{ss} = \left(\frac{Q_{He,ss}}{Q_{He,ss} + Q_{air,ss}} \right) \times 100, \tag{6}$$

where $Q_{He,ss}$ and $Q_{air,ss}$ are the steady-state volume flow rates of helium and air into the garage, respectively. Based on results listed in Table 5, a characteristic quasi-steady-state helium volume percent was 20 %. Substituting this value in Eq. (6) leads to $Q_{air,ss} = 4Q_{He,ss}$. A good estimate for $Q_{He,ss}$ from Table 4 is 4.4×10^{-3} m³/s, which leads to a value of 0.018 m³/s = 63 m³/h for $Q_{air,ss}$. Dividing by the garage volume gives $ACH_{eff,ss} = 0.73$ h^{-1}. This value is roughly a factor of two greater than ACH_{gar} values measured using tracer gases.

The presence of a helium/air mixture is expected to lead to a higher gas exchange rate than when only air is present. Since helium has a lower density than air, its presence inside the garage leads to increased hydrostatic pressure differences between the interior and exterior similar to those created by higher temperatures inside the garage. An average helium volume percent of 20 % creates a density difference equivalent to a temperature difference of 60 °C. This value is much greater than normal temperature differences between the garage interior and either the house or outside. The net result should be a faster exchange rate than when only air is present inside the garage. Another factor is the potential for helium to diffuse rapidly through gypsum drywall. In work described elsewhere, it was demonstrated that such diffusion was a significant removal mechanism for helium from a ¼-scale model of a two-car garage. [5] This was the case even when the drywall was primed and painted, as was the case for the garage in the current study. Due to uncertainties concerning the leak distribution for the garage, it is unclear what the relative roles of the two loss mechanisms were, but it seems likely that both contributed.

Helium volume percent measurements inside the house provide an indication of where a portion of the helium loss from the garage ended up. Since previous work had shown that with the air conditioning fan running, the volume of the entire house was well mixed in about ten minutes [30,31] and the helium releases lasted roughly 4 h, it was a good approximation to assume that any helium entering the house was uniformly mixed in the interior volume. Based on this, it was possible to estimate the helium volume in the house at any given time. Measured helium volume percents in the family room are listed in Table 7 for the six experiments along with estimates for the total helium volume in the house at the end of a release and the corresponding fraction of total helium released present in the house. Recall that results for 9/11/08 and 8/23/10 were considered suspect since a window was opened during the earlier experiment, and the air conditioning was not functioning during the experiment on 8/23/10.

Considerable scatter is evident for the volumes and fractions of released helium present in the house at the end of the release period, with values for the four most reliable tests ranging from 10.9 m^3 to 16.3 m^3 and 0.17 to 0.29, respectively. The relative scatter is much greater than for comparable values included in Table 5 for the garage. The larger scatter suggests that transport of helium from the garage to the house interior and helium losses from the house were more sensitive to some experimental parameter(s) than was the total helium loss from the garage.

The helium volumes present in the house at the end of the releases were comparable (ratios ranging from 0.5 to 1 for the six tests and 0.6 to 1 for the four tests considered more reliable) to those inside the garage. This indicates that anywhere from 30 % to 40 % of the helium loss from the garage during a release in the garage ended up inside the house at the end of a release. This can be compared with the percentage of the wall surface area between the garage and house relative to the total interior surface area for the garage walls and ceiling of 17 %. Presumably, the remainder of the helium was loss to other interior spaces, e.g., attic spaces, and to the outside ambience with lower efficiency.

Next, we focus on helium concentration behavior within the garage during the post-release period. The general concentration behaviors observed following the end of a release were similar for all six experiments, as can be seen in Figure 9, Figure 18, Figure 24, Figure 30, Figure 36, and Figure 45. As evident in Figure 11 for the 9/11/08 experiment, the first effect of halting the helium flow was that helium concentrations at the three lowest sensor locations began to drop almost immediately, while concentrations in the well-mixed upper layer did not immediately respond. As additional time passed, the concentrations at the higher sensor locations began to roll off, with the falloff beginning later for higher sensors. Values of the period required for the helium concentration at the highest sensor to start to fall were summarized in Table 6. Substantial variations are evident. The initial falloff rates were also slower for higher locations.

The roll-off curves in the post-release period were very distinctive. As time passed, helium concentrations at the sensors decayed in such a way that all of the volume percents approached a common value, with measurements at successively higher locations requiring longer periods to reach the common curve. As evident in Figure 75, which shows the helium volume percent time behaviors for sensors located 2.134 m above the floor for the 7/29/10 experiment, there was a distinct change in the helium volume percent falloff rate when the concentration for a given sensor approached the common curve. This change in rate was apparent for upper-layer sensors in all of the experiments. The time required for all of the concentration decays to collapse and the helium levels at this time are listed in Table 6. As discussed earlier, comparison of the decay behaviors for the two 2008 tests (see Figure 9 and Figure 18) suggested that the concentration falloff rates were comparable and that differences in the collapse times resulted from the decays collapsing to different concentrations. This conclusion does not explain all of the variations present in Table 6. As an example, the helium volume percents when a single decay developed for the 8/2/10, 8/6/10, and 8/23/10 experiments were nearly identical to those for the 9/11/08 test, but the period required for the volume percent collapse was much longer for the earliest test.

One of the more interesting aspects of the concentration fall off behaviors concerned the observed time dependence at the lowest sensors. For the 2008 experiments, the helium levels at the 0.152 m sensor dropped rapidly immediately following the end of the release from just above 5 % to a value between 3 % and 4 %. After reaching this level, the concentration remained essentially constant until all of the curves for the higher sensors collapsed to this value. Two slightly different behaviors were observed for the repeated releases without a vehicle present in the garage in 2010. These tests had sensors located at 0.051 m and 0.305 m on the vertical array. For the 7/29/10 test the helium concentration at 0.305 m during the immediate post-release period behaved in a similar way to that observed for the lowest sensor at 0.152 m in the 2008 experiments, while the concentration near the floor actually increased as the higher concentration curves were collapsing. During the 8/3/10 run the helium concentration at the 0.051 m height dropped quickly to a nearly constant value close to the concentration present when all of the higher curves collapsed to a single curve, while the measurements at 0.305 m approached the collapsed curve more slowly. These observations suggest that the mixing behavior in the lower part of the garage depended on some external parameter that varied between the two experiments. Such a dependence is also consistent with differences in the plume mixing behavior between the two experiments described earlier. These findings suggest that more intense mixing was taking place in the lower layer during the 7/29/10 experiment.

Helium volume percent decay constants and corresponding ACH_{gar} values calculated for Sensor #4 when concentrations were less than 1 % are included in Table 6. ACH_{gar} ranged from 0.11 h^{-1} to 0.33 h^{-1}. These values are 2.2 to 6.5 times smaller than the ACH_{eff} estimated above for the quasi-steady-state periods at the end of the helium release phase. The higher ACH_{eff} values require that gas exchange rates into and from the garage did not remain constant during the post-release phase. Even so, if the relative changes were the same for all of the experiments, it would have been reasonable to expect an inverse correlation between the times required for the concentration decay curves to collapse to a common curve and the ACH_{gar} values since faster exchange rates should lead to more rapid mixing and a faster decay. As evident from Table 6, this assumption is consistent with the results for the 7/29/10 data, which had the fastest collapse to a common curve and the largest measured ACH_{gar} value. However, this was not the case for the 9/11/08 and 9/12/08 experiments. The longest collapse period was measured for the 9/11/08 data, but this experiment yielded an ACH_{gar} more than twice as large as that for the 9/12/08 data. The extremely low ACH_{gar} value determined from the 9/12/08 data was accompanied by a time of collapse to a single decay curve that was close to the median value for all of the experiments and was similar in magnitude to three of the other experiments. These observations suggest that gas exchange rates not only decreased during the post-release phase but that there were substantial variations in the relative changes between experiments.

As already discussed, a plausible explanation for these observations was suggested by considering the wind behaviors recorded during the 9/11/08 and 9/12/08 experiments. It was argued that faster exchange rates were favored by higher wind speeds and winds from the south as opposed to the north. The lack of wind data in the immediate vicinity of the house limited our ability to further test this conclusion using the 2010 results.

The period required for the helium concentration to become uniform throughout the garage during the post-release phase was on the order of 4 h. Experiments in the ¼-scale garage showed that molecular diffusion was capable of smoothing out concentration gradients over the entire height on timescales of roughly one hour. [1] The similarity of the mixing times involved suggests that molecular diffusion was playing a role in the observed mixing inside the garage during the post-release period.

The distribution of leaks from the garage and their characteristics were unknown. To our knowledge there have been no previous investigations of release of large amounts of buoyant gas into an actual garage attached to a house in which the concentration measurements have been made. For this reason, it is not known if the observations summarized here are characteristic of all garages or will vary substantially from garage to garage.

Our recent NIST Technical Note is the only study of which we are aware that systematically investigated the effects of vent placement and size on helium losses from a volume. [1] These experiments were done with either one or two vents located in one wall of the ¼-scale two-car garage. For a 4 h release with vents near the top and bottom of the wall, a quasi-steady-state helium distribution developed after roughly two hours, similar to the experiments shown here. In contrast, however, there was a uniform concentration layer only near the ceiling, and the quasi-steady-state helium volume percent decreased with height. When the helium release ended, concentrations at each sensor height began to fall immediately. These behaviors were very different than observed in the current experiments. At long times the concentrations of helium became quasi steady, similar to the observations here, but the collapse for different heights was concurrent, not sequential as in the current experiments. The configuration of the two vents in the reduced-scale garage created significant hydrostatic pressure differences between the interior and ambience that generated strong flows out of the enclosure at the top and into the enclosure at the bottom when high interior helium concentrations were present. The result was a strong recirculating interior flow that generated the observed stratification until the helium levels fell sufficiently to reduce the hydrostatic pressure difference. This induced flow was primarily responsible for the observed concentration behaviors during the release and the immediate post-release periods. The observed differences between the reduced- and full-scale experiments provide strong evidence that a similar vent distribution was not present for the garage experiments. Reduced-scale experiments were also performed with single vents in the center of the front wall. The resulting release and post-release concentration distributions and time behavior were totally different than observed in the current experiments.

In the absence of knowledge of the actual leak distribution, it is only possible to speculate about the mixing flows in the garage based on the observed concentration distributions. The experiments show that during a release, a stable uniform layer of mixed helium and air formed in the upper roughly two thirds of the garage, while a lower concentration region with significant mixing between air from outside and the lower-layer gases existed in the approximately lower one third of the volume. The mixing and air incursion suggest that either a pass-through flow containing large amounts of air entered through at least one wall in the lower part of the garage and exited through one or more different walls or that a recirculating flow with air entering near the floor and lower-layer gas mixture exiting near the top of the lower layer had developed in a single wall. The relatively rapid development of a quasi-steady-state concentration distribution suggests that large amounts of air were entering the garage. This conclusion is supported by the relatively high values of $ACH_{eff,ss}$ estimated above for the gas volume exchange rate between the garage interior and its surroundings.

The experiments also indicated that a large fraction of the helium loss from the garage was preferentially transported into the house. While it is possible that gas from the house interior was also flowing back into the garage, it seems more plausible that the make-up flow consisted primarily of air entering through the garage walls exposed to the outside ambience. If this hypothesis is correct, it requires that the differential pressure between the garage and house was negative. Unfortunately, this measurement was not recorded during the experiments.

As mentioned earlier, rapid helium and hydrogen loss from the ¼-scale garage was observed when the front wall was replaced with gypsum drywall. [5] Both bare and primed and painted walls were tested. Priming and painting slowed down, but did not eliminate the diffusion. A fan was used to mix the gases so that the helium volume fraction measurements could be more easily used to derive effective diffusion coefficients for the gases passing through the walls. Additional experiments not described in [5] were

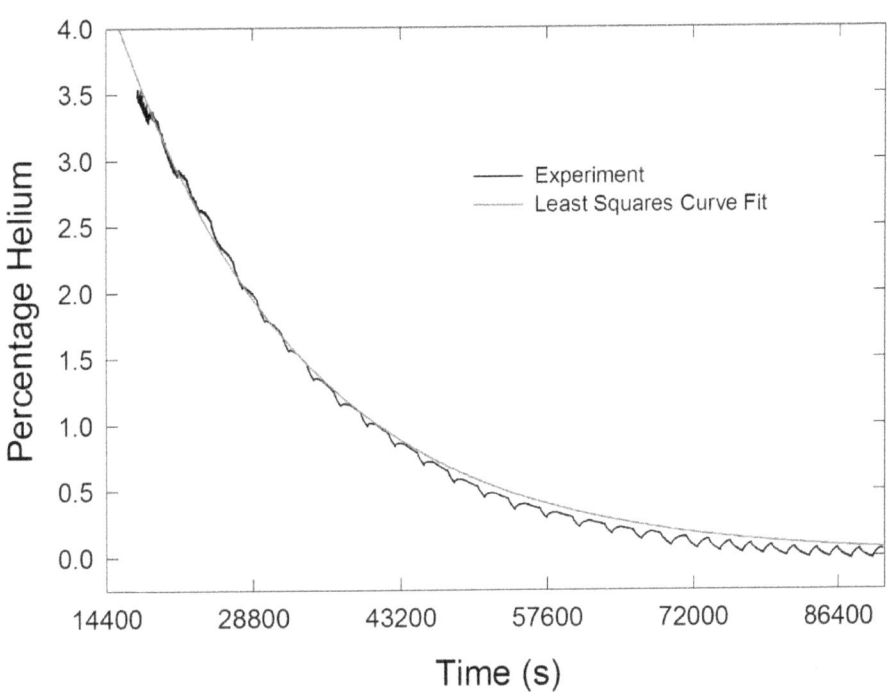

Figure 76. The falloff of helium volume percent in the family room of the house following the end of the helium release inside the garage on 7/29/10 is fit to an exponential curve using a least squares curve fitting procedure.

done without the fan. For these experiments, strong vertical concentration gradients developed, with the largest variations near the ceiling. The absence of such gradients in the current experiments provides evidence that diffusion of helium through the walls did not play a large role in the experiments. It is not currently understood why diffusion should have been less important in the current experiments.

The experiments where helium concentration inside the house was monitored continuously were used to characterize the loss of helium from the house after the helium flow inside the garage was halted. Figure 29 shows data for 7/29/10. In Figure 76 the decay portion of the curve following the end of the helium release has been fit by a least squares procedure assuming the decay was exponential. The resulting time constant was 5.50×10^{-5} s^{-1}, which corresponds to an air change per hour rate for the house, ACH_{hou}, of 0.20 h^{-1}. The R^2 parameter of 0.9974 for the fit indicates the curve was well represented by an exponential. Inspection of Figure 76 revealed that the experimental data fell slightly faster than the calculated fit at long times. The measured ACH_{hou} is well within the range of measurements reported for the house after it was refit. [31]

A similar analysis for the repeat experiment on 8/2/10 yielded a time constant of 7.11×10^{-5} s^{-1}, giving $ACH_{hou} = 0.26$ h^{-1}. This value was slightly higher than for the 7/29/10 experiment. The opposite order was found for the long-time ACH_{gar} values for the garage (see Table 6). The corresponding results for the 8/6/10 data were substantially different, with a time constant of 2.62×10^{-4} s^{-1} and $ACH_{hou} = 0.94$ h^{-1}. The much higher loss rate was due to windows being opened in the house shortly after the helium release was completed. A similar analysis was done for the 8/23/10 experiment even though conditions in the house were likely modified by the absence of air conditioning and fan recirculation until just before the helium flow into the garage was halted. The time constant for the helium volume percent falloff was 8.54×10^{-5} s^{-1}, corresponding to $ACH_{hou} = 0.31$ h^{-1}. These values are similar to those for the 7/29/10 and 8/2/10 experiments and agree with earlier measurements for the house. [31] This suggests that turning on the air conditioning and cooling the house (see Figure 44) had a minimal effect on the house air exchange with its surroundings.

The ACH_{hou} values were very similar in magnitude to corresponding values listed in Table 6 for the garage. This finding reinforces the earlier conclusion that the air tightness of the garage studied here was unusually high, since previous investigations had shown it is not unusual for garages to have ACHs ten times higher than measured for an attached house.

4.2 Garage Experiments with Forced Ventilation

The six experiments employing forced ventilation were performed in 2010 during a period when the central air conditioning for the attached house had malfunctioned, and the whole house fan was not operating. The initial four experiments were done with the Stratus parked over the helium release location, while the Passat was used for the last two. The helium flow and fan exhaust rates for each experiment are listed in Table 8, while parameters used to characterize mixing during the release and post-release periods are included in Table 9.

Four of the experiments (8/12/10, 8/13/10, 8/13/10a, and 8/19/10a) had helium release rates that were between 4.2×10^{-3} m^3/s and 4.3×10^{-3} m^3/s while two (8/9/10 and 8/19/10) had slightly higher rates near 4.43×10^{-3} m^3/s. Four of the experiments were run with exhaust fan settings that generated flows of ≈ 0.092 m^3/s. This value was chosen based on calculations with an analytical model [8] of mixing in the garage in order to limit buildup of helium in the enclosure to volume percents less than the lower flammability limit of hydrogen (4 %) [33]. The two remaining experiments were run with higher exhaust volume flow rates of 0.107 m^3/s, which was the highest flow rate possible with the fan used for the experiment.

The results listed in Table 9 show that the upper-layer concentrations attained quasi-steady-state values of just over 4 % with the lower exhaust flow rates, but that average values along the vertical direction were considerably less than 4 %. When the exhaust flow rate was increased to 0.107 m^3/s, the upper-layer helium volume percent dropped to less than 4 %, and the average vertical concentration fell lower.

The vertical helium volume percent profiles in Figure 49, Figure 55, Figure 59, Figure 63, Figure 67, and Figure 71 show that well-mixed upper layers with higher concentrations than observed near the floor developed in the garage as the helium releases proceeded and extended below the mid height of the garage. After roughly 1800 s of helium flow, the upper-layer concentrations attained a quasi-steady-state behavior in which concentrations grew slowly or leveled off with time. For three of the tests (8/9/10, 8/12/10, 8/19/10a), concentrations for the three lowest sensors located at heights of 0.305 m to 0.914 m were nearly the same, indicating the lower layer was also well mixed. In contrast, the helium volume fraction varied with height over these sensors for the remaining three experiments. For the 8/13/10, 8/13/10a, and 8/19/10 tests, concentration readings for Sensor #2 at 0.610 m were slightly greater than measured for the higher Sensor #3 at 0.914 m, which indicates that a weak density inversion was present in this region of the lower layer. Such inversions are unstable and should result in enhanced mixing. Concentration values recorded by Sensor #4 (1.219 m), which were intermediate between the lower- and upper-layer values, displayed intense fluctuations for the experiments with stratified lower layers, while the helium volume fractions at this sensor height were closer to upper-layer values and had much smaller fluctuations when the lower-layer was well mixed. Whether the intense fluctuations were associated with the density inversions is unknown.

The substantial changes in lower-layer mixing behavior are apparently not correlated with helium release or fan exhaust rates as shown by comparison of the parameters included in Table 8 and Table 9. The lack of weather measurements at the house introduces uncertainties, but comparison with nearby amateur weather stations suggested that the changes were due to weather conditions, with improved mixing in the lower layer associated with higher wind velocities from the south. Wind effects seem to have had little, if any, effects on upper-layer concentrations.

When the helium flow was halted, the volume percents measured at the various sensors began to decrease in a similar manner, but at much faster rates, than observed for the earlier experiments with natural ventilation. Concentrations in the lower layer began to fall almost immediately, while those in the upper layer leveled off for short periods before beginning to fall sequentially from the lower positions to the highest sensor. Eventually, the individual decays collapsed to a common curve. The collapses occurred earlier and for higher helium volume percents for the experiments which had concentration gradients in the lower layer. Two parameters, the required time needed for the helium volume percent to drop to 0.2 % and the decay constant for Sensor #4 at long times, were used to characterize the post-release behavior. The values listed in Table 9 show that both the time period and the falloff rate fell into two groups, with shorter periods and higher falloff rates associated with the higher fan exhaust rate, as expected.

As discussed earlier, the measured decay rates for Sensor #4 were roughly a factor of two higher than the exchange rates that would have been expected based on the exhaust flow rate. It was hypothesized that this was due to the concentration gradients present that resulted in removal of upper-layer gas having helium volume percents greater than the average value.

Helium volume percent measurements were made in the family room of the attached house for the 8/9/10 and 8/12/10 experiments. In both cases, small increases on the order of 0.2 % were observed. These values are not very different from values observed in the house at comparable times for the 8/6/10 (0.07 %, 0.32 %) and 8/23/10 (0 %, 0.15 %) experiments with natural garage ventilation at flow times (1378 s, 2902 s) corresponding to the ends the 8/9/10 and 8/12/10 tests. It should be kept in mind that the house air conditioning fan was not operated on 8/9/10 and 8/12/10, and the helium may not have been fully mixed throughout the house. Even so, the similarity of the helium concentrations is somewhat surprising since the forced ventilation removed upper layer helium from inside the garage and transported it outside. This should have limited helium entering the house in two ways. First, the helium concentrations available for transport into the house were lower, and, second, the fan would have lowered the pressure in the garage somewhat relative to the house, which should have slowed exchange between the garage and the house.

The use of an exhaust fan was effective in limiting the helium volume percents to levels corresponding to hydrogen/air mixtures that would have been too lean to burn. The simple approach described in reference [8] for estimating the required exhaust rate was effective. It should be kept in mind that the ability to limit the buildup of hydrogen depends on both the hydrogen release rate and the fan exhaust rate. The design of an appropriate ventilation system for residential garages will depend critically on the accident scenario chosen to estimate possible hydrogen release rates.

4.3 Buoyant Gas Trapping within Conventional Automobiles

The 8/6/10 and 8/23/10 experiments were done with a vehicle parked over the helium release location and natural ventilation of the garage. Helium volume fraction measurements inside the vehicle provide insights as to how helium concentrations built up inside the automobile during the helium releases and decayed once the helium flow stopped. The relevant concentration measurements were shown in Figure 41 and Figure 46. A plot of the early helium concentration development was included in Figure 42.

The results showed that helium levels built up very rapidly under the vehicle and within the upper part of the engine compartment when helium started to flow. Similar rapid hydrogen concentration growth in a vehicle engine compartment was reported by Maeda et al. for releases under a vehicle located in the open, confirming that hydrogen and helium behave similarly in this type of experiment. [14,15] Measurements inside the upper regions of engine compartments of several different conventional vehicles during hydrogen releases inside a garage at Southwest Research Institute showed rapid increases to similar levels of hydrogen volume percent. [6,7] The current measurements near the bottom of the engine compartment showed much lower concentrations that were close to those measured outside of the vehicle, indicating that the helium concentrations inside the engine compartment were stratified. Maeda et al. reported similar observations for their experiments. [14,15]

The initial helium concentrations formed in the volume of the undercarriage and in the upper part of the engine compartment indicated significant mixing of helium with air as the flow traveled from the release location. The observation that the concentrations continued to increase as helium built up in the surrounding garage provided additional evidence of the importance of this mixing. Additional support for the role of released helium mixing with surrounding gases in the garage is provided by the findings of Maeda et al. [14,15] who showed that, for experiments performed in the open where a buildup of hydrogen in the surroundings was not possible, trapped concentrations of hydrogen attained quasi-steady-state values after only a few seconds and the Southwest Research Institute study, which showed that trapped hydrogen in vehicles inside a garage continued to slowly increase [6,7].

To our knowledge, the only earlier study that has reported the penetration of released gas into passenger compartments of vehicles is the Southwest Research Institute study. [6,7] A number of different conventional automobiles were used. The measurements showed that the growth of hydrogen concentration inside the passenger compartments was much slower than observed inside the engine compartments. There were significant variations in the rate at which hydrogen concentrations increased inside the passenger compartments, suggesting that there were significant vehicle-to-vehicle differences in the degree to which passenger compartments were isolated from the outside environment. It should be kept in mind that the Southwest Research Institute study used older model vehicles that had been scrapped. Such vehicles may have passenger compartments that are leakier than more recent models that are in road condition.

No previous studies were identified in which buoyant gas penetration into the trunk of an automobile was measured. The natural-ventilation experiments revealed that the helium concentration in the passenger compartment (see Figure 41 and Figure 46) rose slightly faster than in the trunk during the early portion of the release period, but helium levels were nearly identical by the end of the 4 h release period. Maximum helium volume percents were just over 16 % for the 8/6/10 and 8/23/10 data. Note that even though the helium concentrations increased relatively slowly over the 4 h release period, they attained values that would have been highly flammable if hydrogen had been released instead of helium. During the post-release phase the helium concentrations in the two compartments decayed relatively slowly and had values which were nearly indistinguishable.

The relatively slow buildup and decay of helium inside the passenger compartment of the Stratus implies a small number of air changes per hour, here denoted ACH_{veh}. In order to estimate ACH_{veh}, an exponential curve was least squares curve fit to the passenger compartment concentration decay at long times. Figure 77 shows the result for the 8/6/10 experiment. The time constant from the fit was 1.87×10^{-5} s^{-1} corresponding to $ACH_{veh} = 0.067$ h^{-1}. As expected, this value is nearly 4 times smaller than values determined for the house and garage. The experimental data for 8/23/10 were not collected for as long, and it was only possible to make a similar calculation as the helium volume percent fell from 2.1 % to 1.4 %. Nonetheless, the resulting value of $ACH_{veh} = 0.088$ h^{-1} was only slightly higher than estimated for the 8/6/10 experiment.

Unlike the current investigation in which the helium penetrated the passenger compartment and trunk from the outside, Ekoto et al. considered the case where hydrogen or helium was injected directly into the passenger compartment or trunk of a conventional automobile. [18] These authors observed that when helium was released directly into the passenger compartment, there was a rapid exchange with the trunk, and concentrations there were slightly lower but tracked those inside the passenger compartment. Interestingly, when the release was into the trunk, the concentration in the passenger compartment did not increase nearly as rapidly as in the trunk. The authors attributed this observation to the effects of stratification and the resulting locations of the interface between the upper and lower layers in the passenger compartment and garage. The rapid attainment of pseudo-steady state concentration distributions and the relatively high concentrations trapped in the compartments provide evidence that the volume exchange rate between the vehicle compartments and outside surroundings was also relatively slow for the vehicle used in their study.

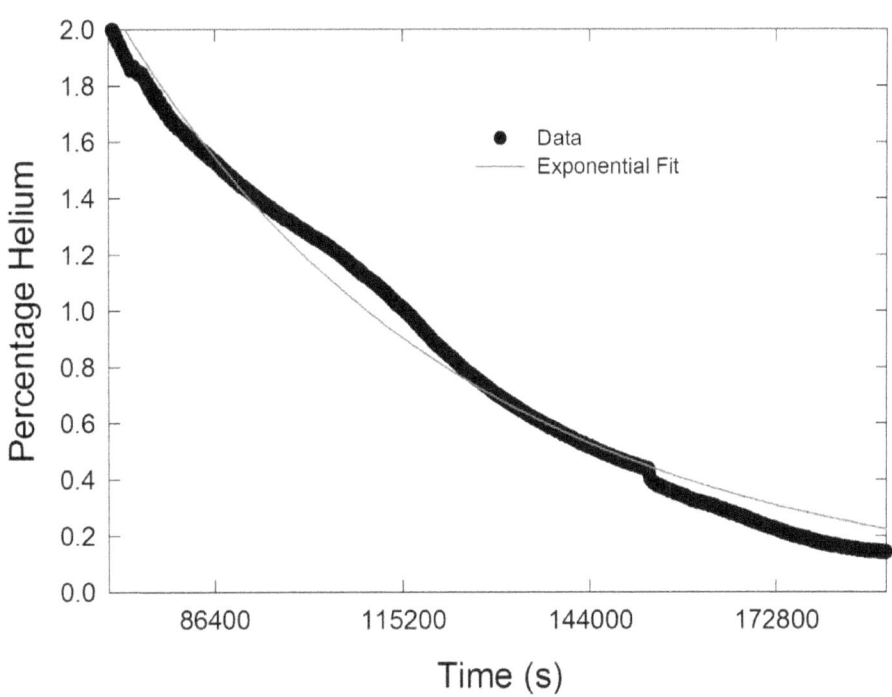

Figure 77. The falloff of helium volume percent in the passenger compartment of the vehicle long after the end of the helium release under the vehicle on 8/6/10 is fit to an exponential curve using a least squares curve fitting procedure.

Six additional experiments with forced ventilation were run with vehicles present in the garage. Four of these were done with the Stratus parked over the release location. As can be seen by comparing the concentration time profiles for locations in the Stratus shown in Figure 41 and Figure 46 for the natural ventilation cases and in Figure 52, Figure 56, Figure 62, and Figure 66 for the experiments with forced ventilation, it is evident the helium volume percent time profiles were nearly identical at short times after the helium flow was initiated. Any minor changes in concentration levels between experiments were most likely associated with small variations in the helium volume flow rates. Recall that for the forced-ventilation experiments, the fan was started within 2 min after the helium flow started. The similarity of the time profiles at short times and the lack of changes in the profiles when the fan flow started show that the garage exhaust flow had little or no direct effect on mixing under the vehicle.

Large differences in the time profiles between the natural- and forced-exhaust experiments appeared as the helium releases continued. For the experiments without forced ventilation, helium volume percents at locations where helium was trapped under the automobile, i.e., in the upper engine compartment, driver's side front wheel well, and undercarriage, continued to increase more slowly after the sharp increases associated with starting the helium flow. These increases were on the order of 25 % at the top of the engine compartment and 75 % in the wheel well and inside the undercarriage at the end of the releases. Helium concentrations in the passenger compartment, trunk, and in the lower part of the engine compartment grew slowly to levels around 15 %. In contrast, during the forced-ventilation experiments, the increases in helium concentration in the trapped areas were much smaller, ≈ 7 % in the upper engine compartment and ≈ 15 % in the wheel well and undercarriage. Concentrations in the passenger compartment and trunk were ≈ 1.5 %, while a value of ≈ 2.5 % was present in the lower engine compartment.

The differences in helium volume percent time development for the two types of ventilation confirm and emphasize the importance of mixing of the surrounding gas with released helium under a vehicle on the levels of helium trapped under and inside a vehicle during a release. The ultimate concentration levels were reduced significantly by using forced ventilation to limit the buildup of helium in the lower layer of the garage. It should be recognized that the helium volume fractions measured under the car in the undercarriage, engine compartment and wheel well correspond to hydrogen concentrations that would be highly flammable. On the positive side, helium levels inside the passenger compartment and trunk were reduced below the flammability limit and would not be expected to ignite if hydrogen was present at the same concentrations.

The final two forced-ventilation experiments were run with the Stratus replaced by the Passat. Comparisons of Figure 62 and Figure 66 (8/13/10 and 8/13/10a) with Figure 70 and Figure 74 (8/19/10 and 8/19/10a) reveal that the general appearances of the concentration time profiles inside the two vehicles were similar. As discussed earlier, concentrations recorded in the undercarriage were not directly comparable since the sensors were located in different areas due to the need to avoid the splash shields present on the Passat. Helium volume percents recorded at the end of a release were somewhat higher in the upper engine compartment and wheel well for the Passat. This may have been due to reduced mixing with surrounding gas as the helium was transported from the release location through the narrow channel present in the undercarriage between the splash guards. Helium levels inside the passenger compartment and trunk of the Passat were about a factor of two higher than observed for the Stratus.

Maeda et al. investigated the build up of hydrogen in the engine compartment of a front-engine, rear-drive unnamed passenger sedan. [15] Concentrations in the upper part of the engine compartment were shown to be nearly uniform. For releases under the center of the vehicle and rates similar to those used in the current work, the quasi-steady-state helium volume percent reported by Maeda et al. was very close to 16 %. This value is similar to those observed following the initial concentration increases during

88

experiments reported here using the Stratus. When the Passat was used, the upper engine compartment helium concentrations were slightly higher. The agreement of experiments using three different vehicles suggests that such concentration levels may be characteristic of mid-size sedans.

Maeda et al. investigated the effect of a vehicle's underside configuration by placing a flat sheet metal cover over the drive-train "tunnel" of their test automobile. [15] Quasi-steady-state concentrations in the engine compartment were reduced by roughly a factor two. A similar affect might have been expected for the Passat, since the undercarriage was partially covered with flat splash shields. However, as noted earlier, there was a deep channel in the undercarriage that passed over the release location between the areas covered with shields. As emphasized by Maeda et al., helium/air mixtures likely flowed along this channel and were conducted into the engine compartment. [15]

The hydrogen trapped under and inside vehicles during releases has been shown to play an important role if hydrogen is subsequently ignited. Studies at SRI International [16] and Southwest Research Institute [6,7] showed that when hydrogen trapped inside a garage with a vehicle parked inside was ignited, the resulting explosions were much more vigorous and damaging than for comparable experiments without a vehicle present. This is likely due to two augmenting effects associated with the presence of a vehicle. The first is the buildup of higher hydrogen concentrations in interior spaces of the vehicle as compared to the garage volume outside. Unless these mixtures become rich, i.e. reach a hydrogen concentration higher than the stoichiometric value, the higher concentration hydrogen trapped inside a vehicle will develop higher flame speeds when ignited. The second effect involves the role of the vehicle itself in accelerating flame speeds. The trapped hydrogen in vehicles is typically found in small partially enclosed spaces containing a great deal of clutter. These conditions are known to accelerate turbulent burning rates and result in stronger explosions than characteristic of comparable hydrogen/air mixtures in the open. [33] For the SRI experiments, it is likely that only the second effect played a role since the hydrogen was released outside of the vehicle, and it is difficult to identify a mechanism whereby higher hydrogen concentrations than present inside the garage would have been present inside the vehicle. In the case of the Southwest Research Institute study, both effects were present since the hydrogen was released under the vehicle and measured concentrations in the engine compartment were considerably higher than in the surrounding garage at the time of ignition. As observed in the current experiments, it is likely that high hydrogen concentrations were also trapped inside the undercarriages of the vehicles used in the Southwest Research Institute investigation.

5 Summary

Twelve experiments have been described in which helium, serving as a surrogate for hydrogen, was released into a residential garage attached to a house at rates between 4.21×10^{-3} m^3/s and 4.44×10^{-3} m^3/s. Helium was released as either momentum- or buoyancy-dominated flow. Cases with natural and forced ventilation of the garage were studied, and the garage was either empty or had a conventional vehicle parked over the release location. Helium concentrations were tracked at multiple locations inside the garage and vehicles (if present). The helium concentration at a single location inside the house was also monitored. Additional measurements included temperatures inside the garage and house and outside and wind velocity for some of the experiments.

Thirty major findings and observations are listed here:
1. The garage had $ACH_{4Pa} \approx 3$ h^{-1} based on doorway blower measurements, but direct measurements using tracer gas methods gave values of ACH_{gar} on the order of 0.4 h^{-1}. The difference in the two values demonstrates that $\Delta P = 4$ Pa, commonly taken as a characteristic pressure difference between building interiors and the outside, is a considerable overestimate. A value closer to $\Delta P = 1$ Pa would be more representative.
2. The air tightness of the test garage was better than typical attached garages in the United States.
3. The leak sizes and their locations were not characterized for the garage.
4. Four hour helium releases with momentum- or buoyancy-dominated flow into the empty garage with natural ventilation yielded similar time profiles of concentration. Quasi-steady-state

vertical helium concentration distributions developed after $\approx$ 2 h that had nearly uniform upper layers extending down to between 1.0 m and 1.2 m above the floor and lower layers with concentrations that fell with decreasing height.

5. Measurements indicated that lateral helium concentration gradients were minimal in the empty garage.

6. Literature findings suggesting more complete mixing with momentum-dominated releases and more concentration stratification with buoyancy-dominated releases were not confirmed by the results in Item 4. It is hypothesized that the reason for the discrepancy is that buoyancy-dominated flows with extremely low Fr mix rapidly and result in reduced stratification and more complete mixing than occurs for buoyancy-dominated flows with Fr closer to the transition value between momentum- and buoyancy-dominated flows.

7. Wind velocity and direction were found to have small, but measurable effects on the rate of exchange between the empty garage interior and its surroundings for natural-ventilation experiments. Higher wind speeds and winds from the south increased the exchange rate. Wind effects were generally small since they were overwhelmed by the larger pressure differences resulting from the large density difference between helium/air mixtures and air.

8. Temperature differences between the garage and house and the garage and outside were generally no larger than a few °C, and no clear correlations of mixing behavior with temperature were observed.

9. Roughly two-thirds of the helium released into the naturally ventilated garage was lost from the garage during the 4 h release period, and between 15 % and 30 % of the released helium was present in the attached house at the end of a release.

10. The maximum helium volume percent in the house for a 4 h experiment with natural ventilation reached just above 4 %, which is close to the lower flammability limit for hydrogen/air mixtures. This result is likely specific to this house and garage combination. In general, helium concentration buildup inside an attached house would be expected to vary with a range of parameters including house volume, interior mixing efficiency, presence of smaller volumes capable of trapping hydrogen, permeability of the house/wall envelop, and physical conditions in the house and garage.

11. Post-release helium concentrations for experiments in the empty, naturally ventilated garage decayed in such a way that the measurements at each height collapsed to a common decay curve at different times, with the period required for collapse increasing with measurement height. Parameters used to characterize this behavior were the helium volume percent present when the curves collapsed and the period required for the highest sensor to reach this concentration.

12. Values of ACH_{gar} estimated using the tails of the helium concentration decay curves for the naturally ventilated garage tests were between 0.11 h^{-1} and 0.33 h^{-1} and were consistent with earlier measurements using tracer gas approaches in the same garage.

13. Placing a vehicle over the helium release location at the center of the floor improved mixing in the garage with natural ventilation by a small, but measurable amount, as reflected by a lowered interface between the uniform upper layer and the lower layer, higher concentrations and smaller gradients in the lower layer, and higher averaged concentrations at the end of the release.

14. Following helium release initiation into a naturally ventilated garage with a Dodge Stratus parked over the release location, helium volume percents jumped abruptly to values around 30 % in the undercarriage volume and around 15 % in the upper engine compartment and at the top of the driver-side wheel well. These values tracked higher as helium concentrations in the surrounding garage increased and then leveled off as the exterior concentrations reached quasi-steady-state levels. Values measured in the lower engine compartment were close to those found in the surroundings immediately outside the vehicle.

15. For the conditions of Item 14, helium concentrations inside the passenger compartment and trunk began to increase slowly after a delay at the start of the release, ultimately reaching levels of ≈ 15 % by the end of a release.

16. The helium concentrations trapped at the various locations in the vehicle during a release attained levels that would be highly flammable if helium were replaced with hydrogen.

17. Helium concentrations in the undercarriage, upper engine compartment, and wheel well for the conditions of Item 14 dropped abruptly within a few 10s of seconds following cessation of the helium release.

18. Helium concentrations for the conditions of Item 14 in the passenger compartment and trunk decayed slowly after the helium flow stopped. Decay rates at long times for repeated experiments were ACH_{veh} = 0.067 h^{-1} and ACH_{veh} = 0.088 h^{-1}. These values are considerably smaller than similar parameters for the house and garage.

19. Active ventilation of the garage containing a vehicle by withdrawing air with a fan significantly reduced helium volume percents inside the garage.

20. Quasi-steady-state concentration distributions developed in less than an hour for the release phase of experiments in the forced-ventilated garage containing a vehicle.

21. Well-mixed upper layers with an interface located above 1.22 m formed during the release phase for the forced-ventilated garage experiments with a vehicle.

22. Significant variations in the lower-layer helium concentration distributions during helium release were observed between different forced-ventilation experiments with vehicles. For some, the lower layer was well mixed and displayed low concentration fluctuations, while for others, vertical concentration gradients developed with indications of an unstable density distribution and large concentration fluctuations. A tentative conclusion, based on comparison with weather data from amateur weather stations near NIST, associated more completely mixed lower layers with higher winds from the south.

23. Observed average helium volume percents at the end of the release for fan exhaust rates around 0.92 m^3/s were on the order of 3.4 %, which is slightly lower than predicted a priori using an analytical model described elsewhere. [8]

24. Increasing the fan exhaust rate by ≈ 15 % resulted in a ≈ 25 % drop in the average quasi-steady-state helium concentration.

25. Small differences in trapped helium concentrations inside and under the vehicle were observed for the two vehicles used in the forced-ventilation garage experiments. The helium volume percents for both vehicles were similar to those reported in an earlier published paper. [15] The similarities suggest the current results will be characteristic for a range of mid-sized automobiles.

26. Helium concentrations present in the engine compartment and trunk of a conventional automobile were reduced at the end of a helium release by forced ventilation since the exterior helium levels surrounding the compartments were reduced.

27. During the post-release period for the forced-ventilation experiments with a vehicle, the vertical concentration distributions developed a stratified structure with a similar appearance to those observed during natural-ventilation tests, but with a much faster falloff.

28. Helium falloff rates at low helium concentrations in the forced-ventilation experiments with a vehicle based on the sensor at 1.22 m were roughly a factor of two higher than predicted assuming instantaneous mixing of entering air with the garage contents. This was attributed to the finite mixing time that led to observed vertical concentration gradients.

29. For the forced-ventilation experiments with a vehicle in the garage, the long term helium falloff rates depended linearly on the fan exhaust rate.

30. During the forced-ventilation experiments with a vehicle in the garage, observed helium volume percents in the family room of the attached house were similar to those measured at comparable times for experiments without forced ventilation. Since the helium release periods for the

forced-ventilation tests were shorter, it is not known if the higher levels observed in 4 h releases with natural ventilation would have been attained.

The findings and observations justify the following conclusions concerning the dispersion behavior and mixing of a highly buoyant gas when released into a residential garage attached to a single family home.

a. The degree of vertical concentration stratification within the garage depends on the Froude number of the release, with large and small Froude numbers favoring more complete mixing.

b. The development of upper and lower layers depends on the Froude number (favored by Froude numbers close to one) and release location (favored by release near the ceiling).

c. The rate of helium loss from the garage depends on the outside wind velocity and direction, but the dependence is relatively weak due to the dominance of hydrostatic pressure differences generated by the substantial density difference between the helium/air mixtures inside the garage and the air in the surroundings.

d. Helium released inside the garage was found to be transferred preferentially through walls into the attached house as opposed to the outside surroundings.

e. Helium released underneath a conventional vehicle builds up substantial concentrations in the undercarriage and engine compartment due to trapping in these partially enclosed spaces, but the spaces were sufficiently open that helium volume fractions dropped rapidly to levels similar to those in the surrounding garage when a release ended.

f. The passenger compartments and trunks of the conventional automobiles investigated were relatively well isolated from exterior volumes inside the vehicle, and, as a result, the surrounding garage, and helium concentration build ups and fall offs were substantially slower than observed at these exterior locations.

g. Moderate active air ventilation was shown to be effective in limiting helium concentrations inside the garage to levels corresponding to nonflammable hydrogen/air mixtures.

h. Moderate active air ventilation was shown to be ineffective in reducing high helium concentrations trapped in the undercarriage and engine compartments of conventional mid-sized passenger vehicles.

References

[1] W. M. Pitts, J. C. Yang, and M. G. Fernandez, Experimental Characterization of Helium Dispersion in a ¼-Scale Two-Car Residential Garage, NIST TN 1694, National Institute of Standards and Technology, Gaithersburg, MD, March 2011, 78 pp.

[2] W. M. Pitts, K. Prasad, J. C. Yang, and M. G. Fernandez, Experimental characterization and modeling of helium dispersion in a ¼-scale two-car residential garage, in Third International Conference on Hydrogen Safety, Ajaccio, France (September 16-18, 2009).

[3] http://www.nist.gov/el/fire_protection/buildings/upload/HeliumDispersionDataSets.zip

[4] K. Prasad, T. Cleary, and J. C. Yang, High pressure release and dispersion of hydrogen in a partially enclosed compartment: effect of natural and forced ventilation, submitted for publication in Intl. J. Hydrogen Energy.

[5] J. C. Yang, W. M. Pitts, M. Fernandez, and K. Prasad, Measurements of effective diffusion coefficient of helium and hydrogen through gypsum, in Fourth International Conference on Hydrogen Safety, San Francisco, CA (September 12-14, 2011).

[6] M. Blais and A. Joyce, Hydrogen release and combustion measurements in a full scale garage, NIST GCR 10-929, National Institute of Standards and Technology, Gaithersburg, MD, January 2010, 56 pp.

[7] W. M. Pitts, J. C. Yang, M. Blais, and A. Joyce, Dispersion and burning behavior of hydrogen released in a full-scale residential garage in the presence and absence of conventional automobiles, in Fourth International Conference on Hydrogen Safety, San Francisco, CA (September 12-14, 2011).

[8] K. Prasad, W. M. Pitts, M. Fernandez, and J. C. Yang, Natural and forced ventilation of buoyant gas released in a full-scale garage: comparison of model predictions and experimental data, in Fourth International Conference on Hydrogen Safety, San Francisco, CA (September 12-14, 2011).

[9] T. G. Cleary and E. L. Johnsson, Detection of hydrogen released in a full-scale residential garage, in Fourth International Conference on Hydrogen Safety, San Francisco, CA (September 12-14, 2011).

[10] K. Prasad, and J. Yang, Effect of wind and buoyancy on hydrogen release and dispersion in a compartment with vents at multiple levels, Intl. J. Hydrogen Energy **35** (1), 9218–9231 (2010).

[11] K. R. Prasad and J. C. Yang, Vertical release of hydrogen in a partially enclosed compartment: role of wind and buoyancy, Intl. J. Hydrogen Energy **36** (1), 2489–2496 (2011).

[12] K. Prasad, W. M. Pitts, and J. C. Yang, A numerical study of the release and dispersion of a buoyant gas in partially confined spaces, Intl. J. Hydrogen Energy **36** (8), 5200–5210 (2011).

[13] B. Cariteau, J. Brinster, E. Studer, I. Tkatschenko, and G. Joncquet, Experimental results on the dispersion of buoyant gas in a full scale garage from a complex source, Intl. J. Hydrogen Energy **36** (3), 1094–1106 (2011).

[14] Y. Maeda, M. Takahashi, Y. Tamura, J. Suzuki, and S. Watanabe, Test of vehicle ignition due to hydrogen gas leakage, SAE Technical Paper Series, SP-1990 Fire Safety, 1-7 (2006).

[15] Y. Maeda, H. Itoi, Y. Tamura, J. Suzuki, and S. Watanabe, Diffusion and ignition behavior on the assumption of hydrogen leakage from a hydrogen-fueled Vehicle, SAE 2007 Transactions, Journal of Passenger Car: Mechanical System, Section 6 **116**, 233–239 (2007).

[16] E. G. Merilo, M. A. Groethe, J. D. Colton, and S. Chib, Experimental study of hydrogen release accidents in a vehicle garage, Intl. J. Hydrogen Energy **36** (3), 2436-2444 (2011).

[17] 2002 ICC Final Action Agenda, M7-02, 304.3, Section 1307.4, p. 461. Available from: www.iccsafe.org/cs/codes/2002cycle/faa02imc.PDF

[18] I. W. Ekoto, E. G. Merilo, D. E. Dedrick, and M. A. Groethe, Performance-based testing for hydrogen leakage into passenger vehicle compartments, Intl. J. Hydrogen Energy **36** (16), 10169-10178 (2011).

[19] H. Liu and W. Schreiber, The effect of ventilation system design on hydrogen dispersion in a sedan, Intl. J. Hydrogen Energy **33** (19), 5115-5119 (2008).

[20] M. H. Sherman and R. Chan, Building airtightness: research and practice, Report LBNL-53356, Lawrence Berkeley National Laboratory, Livermore, CA, 2003, 46 pp.

[21] W. R. Chan, P. N. Price, and A. J. Gadgil, Sheltering in buildings from large-scale outdoor releases, Report LBNL-55575, Lawrence Berkeley National Laboratory, Livermore, CA, July 2004, 10 pp.

[22] S. J. Emmerich, J. E. Gorfain, and J. E. Howard-Reed, Air and pollutant transport from attached garages to residential living spaces – literature review and field tests, Intl. J. Vent. **2** (3), 265-276 (2003).

[23] American Society of Heating, Refrigeration, and Air Conditioning Engineers. Ventilation for acceptable indoor air quality, Atlanta (GA): ANSI/ASHRAE Standard 62-1989; 1990.

[24] M. R. Swain, J. Shriber, and M. N. Swain, Comparisons of hydrogen, natural gas, liquidfied petroleum gas, and gasoline leakage in a residential garage, Energy Fuels **12** (1), 83-99 (1998)

[25] International Code Council, Inc. International mechanical code. Country Club Hills (IL) (2009).

[26] Garage performance testing, Research Highlight, Technical Series 04-108, Canada Mortgage and Housing Corporation Ottawa, Ontario, Canada, April 2004, 4 pp.

[27] S. Batterman, G. Hatzivasilis, and C. Jia, Concentrations and emissions of gasoline and other vapors from residential vehicle garages, Atmos Environ **40** (10), 1828-1844 (2006).

[28] L. R. Waterland, C. Powars, and P. Stickles, Safety evaluation of the FuelMaker home refueling concept. Subcontractor Report: NREL/SR-540-36780, National Renewable Energy Laboratory; Golden, CO February 2005, 190 pp.

[29] P. Adams, A Bengaouer, B. Cariteau, V. Molkov, and A. G. Venetsanos, Allowable hydrogen permeation rate from road vehicles, Intl. J. Hydrogen Energy **36** (3), 2742-2749 (2011).

[30] S. Nabinger and A. Persily, Airtightness, ventilation, and energy consumption in a manufactured house: Pre-retrofit results, NISTIR 7478, National Institute of Standards and Technology, Gaithersburg, MD, May 2008, 31 pp.

[31] S. Nabinger, A. Persily, and W. S. Dols, Impacts of airtightening retrofits on ventilation rates and energy consumption in a manufactured home, NISTTN 1673, National Institute of Standards and Technology, Gaithersburg, MD, October 2010, 30 pp.

[32] C. J. Chen and W. Rodi, Vertical turbulent buoyant jets—A review of experimental data, 1980, Pergamon Press, Oxford.

[33] Hydrogen Fundamentals, Biennial Report on Hydrogen Safety (Version 1.2), The International Association for Hydrogen Safety, http://www.hysafe.org/BRHS, 2005.